T0199860

Elastic Optical Networks:
Fundamentals, Design, Control, and Management

Bijoy Chand Chatterjee
Department of Computer Science
South Asian University, New Delhi, India

Eiji Oki
Graduate School of Informatics
Kyoto University, Kyoto, Japan

CRC Press
Taylor & Francis Group
Boca Raton   London   New York

CRC Press is an imprint of the
Taylor & Francis Group, an **Informa** business

A SCIENCE PUBLISHERS BOOK

CRC Press
Taylor & Francis Group
6000 Broken Sound Parkway NW, Suite 300
Boca Raton, FL 33487-2742

© 2020 by Taylor & Francis Group, LLC
CRC Press is an imprint of Taylor & Francis Group, an Informa business

No claim to original U.S. Government works

Version Date: 20200229

International Standard Book Number-13: 978-1-138-61171-9 (Hardback)
International Standard Book Number-13: 978-0-367-51021-3 (Paperback)

This book contains information obtained from authentic and highly regarded sources. Reasonable efforts have been made to publish reliable data and information, but the author and publisher cannot assume responsibility for the validity of all materials or the consequences of their use. The authors and publishers have attempted to trace the copyright holders of all material reproduced in this publication and apologize to copyright holders if permission to publish in this form has not been obtained. If any copyright material has not been acknowledged please write and let us know so we may rectify in any future reprint.

Except as permitted under U.S. Copyright Law, no part of this book may be reprinted, reproduced, transmitted, or utilized in any form by any electronic, mechanical, or other means, now known or hereafter invented, including photocopying, microfilming, and recording, or in any information storage or retrieval system, without written permission from the publishers.

For permission to photocopy or use material electronically from this work, please access www.copyright.com (http://www.copyright.com/) or contact the Copyright Clearance Center, Inc. (CCC), 222 Rosewood Drive, Danvers, MA 01923, 978-750-8400. CCC is a not-for-profit organization that provides licenses and registration for a variety of users. For organizations that have been granted a photocopy license by the CCC, a separate system of payment has been arranged.

Trademark Notice: Product or corporate names may be trademarks or registered trademarks, and are used only for identification and explanation without intent to infringe.

Visit the Taylor & Francis Web site at
http://www.taylorandfrancis.com

and the CRC Press Web site at
http://www.crcpress.com

Preface

The rapid growth in worldwide communications and the brisk adoption of the Internet have significantly modified our way of life. This revolution has led to a vast growth in communication bandwidth in every year. An optical network has the potential to support the continued demands for communication bandwidth. Unfortunately, a conventional optical network is incapable of achieving the enormous bandwidth demanded by clients as it is fatally hindered by the electrical bandwidth barrier. Flexgrid technology is now considered to be a promising solution for future high-speed network design. To promote an efficient and scalable implementation of elastic optical technology in the telecommunications infrastructure, many challenging issues related to routing and spectrum allocation (RSA), resource utilization, fault management and quality of service provisioning must be addressed with utmost importance. This book makes key contributions to the development of elastic optical networks (EONs), concerning RSA problems with spectrum fragment issues, which degrade the quality of service provisioning. The contributions made by this book are explained as follows.

Beginning with a brief introduction to the optical fiber transmission system, this book then moves into an overview of the wavelength division multiplexing (WDM), and WDM optical networks. It discusses the limitations of conventional WDM optical networks, and provides examples of how elastic optical networks overcome these limitations. It then presents the architecture of the elastic optical network and its operation principle. To complete the discussion of network architecture, this book focuses on the different node architectures, and compares their performance in terms of scalability and flexibility. This book reviews and classifies different RSA approaches, including their pros and cons. It focuses on different aspects related to RSA. The spectrum fragmentation is a serious issue in EONs, which is needed to manage it efficiently. This book explains the fragmentation problem in EONs, discusses, and analyzes the major conventional spectrum allocation policies in terms of the fragmentation effect in a network.

The taxonomies of the fragmentation management approaches are presented along with different node architectures. Subsequently, this book reviews state-of-the-art fragmentation management approaches.

An interesting part of this book, is that it provides mathematical modeling and analyzes theoretical computational complexity for different problems in elastic optical networks. Finally, it addresses the research challenges and open issues in EONs and provides future directions for future research.

There are several books available in the area of EONs, which focus on the architecture, technologies, control, and management aspects. However, there is a gap between the theory and its practical implementation. This book is intended to fill this gap and provides fundamentals and advanced concepts for elastic optical networking industries, including both practical and theoretical aspects. We believe that this book will serve as a useful addition to the literature. It not only describes fundamental and theoretical aspects, but also provides a practical guide to the understanding of network control and design using mathematical modeling and algorithms in the domain of EONs.

A graduate course on optical networking has been offered by Bijoy Chand Chatterjee at South Asian University, New Delhi, India, and some of the contents of this book have been used for this course. Eiji Oki used a draft of this book as the text for gradate and undergraduate courses that he taught on communication networks, information networks, and optical communications at Kyoto University, Kyoto, Japan and The University of Electro-Communications, Tokyo, Japan, and has kept improving the draft based on student feedback since 2013. These courses continue to attract both academic and industrial practitioners, as design and control for optical networks are one of the key topics in the information and communication technology industry.

This book primarily targeted at both senior and graduate students who are interested in advanced technology in elastic optical networks (EONs). It is meant for senior and graduate students in electrical engineering, computer engineering, and computer science. Using this book, students will understand both fundamental and advanced technologies in EONs. It provides mathematical modeling and analysis of theoretical computational complexity for different problems in EONs and fundamental and advanced concepts for elastic optical networking industries including both practical and theoretical aspects.

This book is also intended for optical networking professionals, R&D engineers, and network designers, who are currently active or anticipate the future development of optical networks. This book will allow them to design a cost-effective optical network while improving the network performances including call blocking, quality-of-service, and reliability.

The minimum requirements to understand this book is a knowledge of algorithms, and the fundamental concepts of optical networks. Some background in communication networks would be useful. All the concepts in this book are developed from intuitive basics, with further insight provided through examples of practical applications.

Organization

The book is organized as follows.

■ Chapter 1 briefly discuses the WDM based optical networks and the shortcomings of current technology.

■ Chapter 2 presents the motivation of the EON and introduces its unique concepts and enabling technology.

■ To fulfill our ever-increasing bandwidth demands, EONs are indispensable. The performance of the EON depends on its network and node architectures. Chapter 3 presents the architecture of the EON and its operation principle.

■ RSA in EONs is considered one of the key functionalities due to its information transparency and spectrum reuse characteristics. Chapter 4 reviews and classifies RSA approaches, including their pros and cons, and various issues related to RSA are discussed in Chapter 5.

■ Chapter 6 presents the fragmentation problem in EONs and discusses the different metrics used to measure the spectrum fragmentation for an EON link, and compares them in terms of their pros and cons. This chapter also presents how to estimate the fragmentation of an entire network. The major spectrum allocation approaches, which are random fit, last fit, first fit, first-last fit, least used, most used, and exact fit, are analyzed in terms of overall network fragmentation.

■ Chapters 7 and 8 discuss different fragmentation management approaches considering non-defragmentation and defragmentation approaches, respectively. Chapter 9 presents and analyzes different defragmentation schemes in 1+1 path protected EONs.

■ Chapter 10 exploits spectrum fragmentation management in EONs considering software-defined networks (SDN).

■ Chapter 11 starts with a general description of optimization problems, and then presents different integer linear programming formulation for problems related to EONs.

■ Chapter 12 starts with a general description of computational complexity analysis of a problem, and then shows the approach how to prove NP-completeness of a problem. Finally, this chapter shows the proof of NP-completeness of some problems related to EONs.

■ Chapter 13 addresses research issues and challenges faced by optical network researchers and shows some directions for further research.

We provide several exercises at the end of each chapter so that readers can check their understanding of the chapter. Finally, we show the answer examples to the exercises of each chapter at the end of the book.

Acknowledgement

This book could not have been published without the help of many people. We thank them for their efforts in improving the quality of the book. We have done our best to accurately describe the basic concepts and are responsible for any remaining error. If any error is found, please send an e-mail to bijoycc@ieee.org and oki@i.kyoto-u.ac.jp. We will correct them in future editions.

Several chapters of the book are based on our research works. We would like to thank our collaborators as follows: Andrea Fumagalli, Fujun He, Naoaki Yamanaka, Nattapong Kitsuwan, Nityananda Sarma, Waya Fadini, Satoru Okamoto, Seydou Ba, Takaaki Sawa, and Takehiro Sato. The manuscript draft was reviewed by Abdul Wadud, Imran Ahmed, and Kenta Takeda. We are immensely grateful for their comments and suggestions.

Bijoy Chand Chatterjee wishes to thank his wife, Satabdi, and his mother, Minati, for their love. Eiji Oki wishes to thank his wife, Naoko, his daughter, Kanako, and his son, Shunji, for their love.

Bijoy Chand Chatterjee
Eiji Oki

Contents

About the Authors

Bijoy Chand Chatterjee received the Ph.D. degree from the Department of Computer Science & Engineering, Tezpur University in the year of 2014. From 2014 to 2017, he was a Postdoctoral Researcher in the Department of Communication Engineering and Informatics, the University of Electro-Communications, Tokyo, Japan, where he was engaged in researching and developing high-speed flexible optical backbone networks. From 2017 to 2018, he worked as a DST Inspire Faculty at Indraprastha Institute of Information Technology Delhi (IIITD), New Delhi, India. Currently, he is working at South Asian University (An international university established by SAARC nations), New Delhi, India, as an Assistant Professor and DST Inspire Faculty. He is also an Adjunct Professor at Indraprastha Institute of Information Technology Delhi (IIITD), New Delhi. Before joining South Asian University, he was an ERCIM Postdoctoral Researcher at the Norwegian University of Science and Technology (NTNU), Norway. He was a Visiting Researcher at Oki Lab, Kyoto University, Japan, from Jun-July, 2019. Dr. Chatterjee was the recipient of several prestigious awards, including DST Inspire Faculty Award in 2017, ERCIM Postdoctoral Research Fellowship by the European Research Consortium for Informatics and Mathematics in 2016, UEC Postdoctoral Research Fellowship by the University of Electro-Communications, Tokyo, Japan in 2014, and IETE Research Fellowship by the Institution of Electronics and Telecommunication Engineers, India in 2011. His research interests include QoS-aware protocols, cross-layer design, fog networking, optical networks and elastic optical networks. He has published more than 50 journal/conference papers. He has authored of the book *Routing and Wavelength Assignment for WDM-based Optical Networks: Quality-of-Service and Fault Resilience*, published by Springer International Publishing, Cham, in 2016. Currently, he has been serving as an associate editor in IEEE Access. He is a Life Member of IETE and a Senior Member of IEEE.

Eiji Oki is a Professor at Kyoto University, Japan. He received the B.E. and M.E. degrees in instrumentation engineering and a Ph.D. degree in electrical engineering from Keio University, Yokohama, Japan, in 1991, 1993, and 1999, respectively. In 1993, he joined Nippon Telegraph and Telephone Corporation (NTT) Communication Switching Laboratories, Tokyo, Japan. He has been researching network design and control, traffic-control methods, and high-speed switching systems. From 2000 to 2001, he was a Visiting Scholar at the Polytechnic Institute of New York University, Brooklyn, New York, where he was involved in designing terabit switch/router systems. From 2008 to 2017, he was with The University of Electro-Communications, Tokyo. He joined Kyoto University, Japan in March 2017. He has been active in standardization of path computation element (PCE) in the IETF. He wrote more than ten IETF RFCs. Prof. Oki was a recipient of several prestigious awards, including the 1999 IEICE Excellent Paper Award, the 2001 Asia-Pacific Outstanding Young Researcher Award presented by IEEE Communications Society for his contribution to broadband network, ATM, and optical IP technologies, the 2010 Telecom System Technology Prize by the Telecommunications Advanced Foundation, IEEE HPSR Paper Awards in 2012, 2014, and 2019, the 2015 IEICE Achievement Award, IEEE Globecom 2015 Best Paper Award, 2016 Fabio Neri Best Paper Award Runner Up, and Excellent Paper Award of 2019 Information and Communication Technology on Convergence. He has authored/co-authored five books, *Broadband Packet Switching Technologies*, published by John Wiley, New York, in 2001, *GMPLS Technologies*, published by CRC Press, Boca Raton, FL, in 2005, *Advanced Internet Protocols, Services, and Applications*, published by Wiley, New York, in 2012, *Linear Programming and Algorithms for Communication Networks*, CRC Press, Boca Raton, FL, in 2012, and *Routing and Wavelength Assignment for WDM-based Optical Networks: Quality-of-Service and Fault Resilience*, Springer, Cham, in 2016. He is a Fellow of IEEE and a Fellow of IEICE.

Chapter 1

Introduction to Optical Networks

After experiencing rapid growth during the late 90s, the telecom industry has been experiencing challenging times, such as a high bandwidth requirement, over the past few years. We need to be ready with the appropriate technology and engineering solutions to meet the growing bandwidth needs of our information society. To fulfill our ever-increasing bandwidth demand, all optical backbone networks along with wavelength-division multiplexing (WDM) technologies are essential, this is due to their many desirable properties including higher bandwidth availability, low signal attenuation, low signal distortion, low power requirement, low material usage, small space requirement, and low cost. This chapter briefly discusses the optical networks and the shortcomings of current technology.

1.1 Overview of telecom networks

An overview of telecommunication networks is depicted in Fig. 1.1, which consists of three major components, namely (i) access network, (ii) metropolitan-area network, and (iii) backbone network.

The access network is mainly responsible for enabling end-users, such as businesses and residential customers, to get connected to the rest of the network infrastructure, which typically spans a few kilometers. The access network continues to be a bottleneck, and users require a higher bandwidth to be delivered to their machines. Several approaches, such as passive optical networks (PONs),

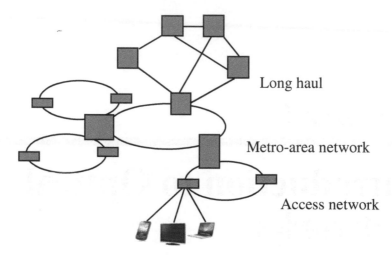

Figure 1.1: Overview of telecom networks.

Ethernet PON technology (EPON) have been considered to offer high bandwidth requirement.

The metro-area network usually spans a metropolitan area and covers anything from 10-100 kilometers with interconnecting access and long-haul networks. Metro networks today are based on synchronous digital hierarchy (SDH)/ synchronous optical network (SONET), as SONET/SDH has been very successful in delivering the early wave of end-user connectivity.

The long-haul network or backbone network spans long distances, and each link of these could be a few hundred to a few thousands kilometers in length. They are established to provide nationwide or global coverage. The long-haul network or backbone network is based on WDM technology.

1.2 Overview of optical networks

An optical network is a typically a data communication network that is made with fiber optics technology. Optical fiber cables are used as the major communication medium in optical networks where data is converted and transmitted as light pulses between sender and receiver nodes.

An optical network is not essentially all-optical; the transmission is surely optical, but the switching can be optical, electrical, or hybrid. An optical network is not essentially packet-switched; it can be a circuit switching based network. It can follow either a fixed grid or a flexible grid.

In a communication service, a circuit is used to connect end-to-end communication terminals. It is mandatory that the circuit can be established at any two

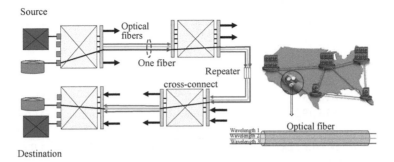

Figure 1.2: Optical path network.

terminals, and hence a circuit network can provide a mesh like connection. If several circuits are directly lodged into a physical communication medium, such as an optical fiber network, a network operator may face difficulties for the operation and management of the circuit network and transmission media network; there exists a large amount of granularity differences among circuits' capacities (kbps order) and transmission line capacities (few hundreds Mbps to Gbps order). To address this granularity mismatch, a path network [1] was introduced between the circuit network and the transmission media layer, as shown in Fig. 1.2. A path is a pack of circuits, which is an accommodation unit in the transmission line. The path network is used to deploy networking technologies, including recovery from transmission line and node failure and traffic engineering.

In the late 1980s, varieties of optical networks, namely, enterprise serial connection [2], fiber distributed data interface [3], token-ring [4], ethernet [5], and SONET/SDH [6, 7] were developed as a replacement of copper cable to achieve a higher communication bandwidth. Among those networks, SONET/SDH had provided the basis for the current high-speed backbone networks. It has also been considered to be one of the most successful standards in the entire networking industry.

SONET/SDH[6, 7] is a standardized protocol that transfers multiple digital bit streams over optical fiber using lasers or highly coherent light from light-emitting diodes (LEDs). It provides support for the operations, administration, and maintenance (OAM) functions that are required to operate digital transmission facilities. SONET has defined a hierarchy of signals called synchronous transport signals (STSs). These levels are known as synchronous transport modules (STMs). The physical links that transmit each level of STS are called optical carriers (OCs). The optical carrier equivalent to STS-1 is OC-1, which supports a data rate of 51.84 Mb/s. Table 1.1 provides the hierarchy of the most common SONET/SDH data rates. A typical SONET transmission system consists of a transmission path and devices as depicted in Fig. 1.3. In this figure, STS multiplexers and demultiplexers perform the task of multiplexing several incom-

Table 1.1: SONET/SDH digital hierarchy.

OL	EL	LR	PR	OR	SE
OC-1	STS-1	51.840	50.112	1.728	-
OC-3	STS-3	155.520	150.336	5.184	STM-1
OC-12	STS-12	622.080	601.344	20.736	STM-4
OC-48	STS-48	2488.320	2405.376	82.944	STM-16
OC-192	STS-192	9953.280	9621.504	331.776	STM-64
OC-768	STS-768	39813.120	38486.016	1327.104	STM-256

OL: Optical Level; EL: Electrical Level; LR: Line Rate (Mbps); PR: Payload Rate (Mbps); OR: Overhead Rate (Mbps); SE: SDH Equivalent

ing signals onto single trunk and vice versa. Add-drop multiplexers are used in SONET technology to add signals and remove a required signal from the data stream without demultiplexing the entire signal. SONET consists of four functional layers, namely, (i) photonic layer, (ii) section layer, (iii) line layer and (iv) path layer. Photonic layer communicates to the physical layer of Open System Interconnection (OSI) model which is concerned with transmission of optical pulses. The section layer deals with signals in their electrical form. It also handles framing, scrambling, and error control. The line layer is concerned with the multiplexing and demultiplexing of signals. The path layer handles the transmission of a signal from source to destination.

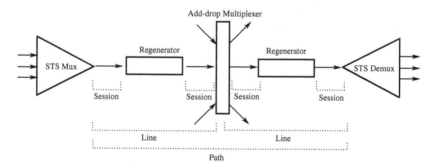

Figure 1.3: Concept of SONET system.

1.2.1   Optical transmission system

A typical optical transmission system has three basic components, namely, (i) transmitter, (ii) transmission medium, and (iii) receiver. The transmitter is used to convert data into a sequence of on/off light pulses. These light pulses are transmitted through the transmission medium and finally, converted back to the original data at the receiver side. An optical transmitter is essentially a light source. Although, initially LEDs had been used as a light source but nowadays, all optical networks use lasers to produce high-powered beams of light. Optical fiber has been used as a transmission medium in optical communication systems. Normally, a photodiode can be used as a receiver to convert a stream of photons (optical signal) into a stream of electrons (electrical signal). It has been observed that when a light pulse propagates through optical fiber, it is distorted. This distortion occurs mainly due to physical layer impairments [8]. Physical layer impairments can be classified into two categories, namely, (i) linear impairments (LIs) and (ii) non-linear impairments (NLIs), which are discussed in the following.

The most important linear impairments for signal distortion are signal attenuation and dispersion. Signal attenuation [9] in fiber leads to loss of signal power due to impurities in the fiber glass and Rayleigh scattering [9]. Signal attenuation (α) is measured in decibels using (1.1).

$$\alpha = -\frac{10}{L}\log_{10}\left(\frac{P_{\text{out}}}{P_{\text{in}}}\right), \tag{1.1}$$

where L, P_{out}, and P_{in} denote fiber loss in km, output power, and input power, respectively.

Figure 1.4 shows the attenuation in decibels per kilometer of fiber for different wavelengths. From the figure, it can be observed that three main low-loss bands are centered at 0.850, 1.300 and 1.550 micron. Among these bands, C band (1.530-1.565 micron) and L band (1.565-1.625 micron) have been usually used to achieve a huge communication bandwidth due to lower attenuation. To overcome attenuation, repeaters are placed to restore the degraded signal for continuing further transmission.

On the other hand, when the light pulses propagate through optical fiber, the pulses spread out (*i.e.,* duration of the pulses broaden). This spreading of light pulses is called dispersion [9]. Dispersions in optical fiber are mainly classified into three categories, namely, (i) material dispersion (MD), (ii) waveguide dispersion (WD) and (iii) polarization mode dispersion (PMD). MD occurs due to the refractive index which varies as a function of the optical wavelength. WD is caused by the wavelength dependence of the group velocity due to specific fiber geometry. It describes the dependence of the effective refraction index on the normalized frequency of radiation propagating through the optical fiber. The waveguide dispersion results in distribution changes of power between the core and the cladding. PMD is a form of modal dispersion, where different polarizations of the optical signal travel with different group velocities due to random imperfections and asymmetries. PMD plays an important role when the bit rate of a channel, which is the type of transmission media that is used to transfer a message from one point to another, is higher than or equal to 10 Gbps [10,21].

The non-linear effects [8] in optical fiber occur either due to change in the refractive index of the medium with optical intensity (power) or due to inelastic-

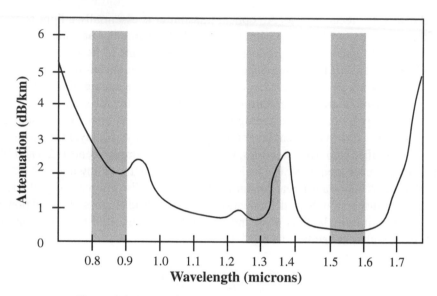

Figure 1.4: Attenuation versus wavelength for optical fiber.

scattering phenomenon. The important non-linear impairments are: (i) self phase modulation (SPM), (ii) four wave mixing (FWM), (iii) cross phase modulation (XPM), (iv) stimulated brillouin scattering (SBS) and (v) stimulated raman scattering (SRS). SPM [11] is a nonlinear optical effect of light-matter interaction. An ultrashort pulse of light, when traveling in a medium, a varying refractive index of the medium is introduced due to the optical Kerr effect [12], which produces a phase shift in the pulse, and leads to a change of the pulse's frequency spectrum. FWM [13] is an intermodulation phenomenon in non-linear optics, which is introduced if at least two different wavelength components propagate together. When the optical power from a wavelength impacts the refractive index, the impact of the new refractive index on another wavelength is known as XPM [14]. SBS [15] refers to the interaction of light with the material waves in a medium, which is facilitated by the refractive index dependence on the material properties of the medium. SRS [16] takes place when an excess of stokes photons that were previously generated by normal Raman scattering are present or are deliberately added to the excitation beam. The detailed descriptions of these non-linear effects can be found in [8, 9].

1.2.2 Wavelength division multiplexing

An optical fiber has an enormous bandwidth capacity, but the accessing rate of the end-user (for example, a workstation) is limited which is a few gigabits per second. Therefore, it is extremely difficult to exploit all the vast communication

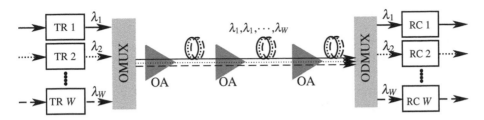

Figure 1.5: Concept of wavelength division multiplexing.

bandwidth of a single fiber using a single wavelength channel due to the optical-electronic bandwidth mismatch. Wavelength division multiplexing (WDM) [6,7] is a technique that manages the opto-electronic bandwidth mismatch by multiplexing wavelengths of different frequencies onto a single fiber channel as shown in Fig. 1.5. WDM creates many virtual fibers and each of them can capable of carrying a different signal. Each signal can be carried at a different rate like - OC-3/STM-1, OC-48/STM-16, and so on and in a different format, such as, SONET/SDH, asynchronous transfer mode (ATM), data, and so on. Therefore, the capacity of existing networks can be improved using the WDM technology, without upgrading the network.

WDM systems are categorized mainly into two different wavelength patterns, which are coarse wavelength division multiplexing (CWDM) and dense wavelength division multiplexing (DWDM). Coarse WDM systems typically use the band from 1.271 micron to 1.611 micron and support up to 17 channels, when 20 nm spacing is used. If Dense WDM (DWDM) systems use the C-Band (1.530 micron to 1.565 micron) transmission window with 100 GHz spacing, approximate 40 channels can be obtained; approximate, 80 channels can be achieved when 50 GHz spacing is used. The spacing between two wavelengths, denoted by $\Delta\lambda$, is estimated by (1.2)

$$\Delta\lambda = \frac{\lambda_0^2}{c} \cdot \Delta f, \tag{1.2}$$

where λ_0, Δf, and c are the center wavelength, frequency spacing, and speed of the light in free space, respectively [17]. At a wavelength $\lambda_0 = 1550$ nm, a wavelength spacing of 0.8 nm corresponds to a frequency spacing of 100 GHz, a typical spacing in WDM systems.

1.2.3 Optical network architecture

The architecture of optical networks are being mainly classified into two categories, namely, (i) broadcast-and-select optical networks and (ii) wavelength-routed optical networks. The following subsections briefly explain these networks.

1.2.3.1 Broadcast-and-select optical networks

Broadcast-and-select optical networks [6, 7, 18] consist of a number of nodes. These nodes are connected through optical fibers to a passive star coupler. Figure 1.6 shows a passive-star-based local optical network. In broadcast-and-select optical networks, nodes are equipped with fixed or tunable transmitters to transmit signals on different wavelengths. These signals are combined into a single signal by the passive star coupler. Then, this combined signal is broadcasted to all the nodes in the network. The power of the transmitted signal is split equally among all the output ports leading to all nodes in the network. Each node can select a required wavelength to receive the desired signal by tuning its receiver to that wavelength. The communication between the transmitter and receiver can be classified into two categories, such as (i) single-hop communication and (ii) multi-hop communication. In single-hop communication, transmitted signals travel from source to destination entirely in optical domain. Multi-hop communication transmits the signal through a certain number of wavelengths and thus forms a virtual path over the physical path.

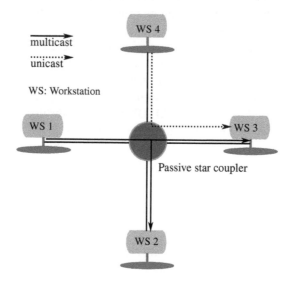

Figure 1.6: A passive-star-based local optical network.

The broadcast-and-select networks can easily support multi-cast traffic. Therefore, multiple receivers at different nodes can be tuned to receive the same wavelength. The main drawbacks of broadcast-and-select network are as follows—(i) it requires synchronization and rapid tuning, (ii) it cannot support wavelength reuse characteristics and hence a large number of wavelength channels is required, (iii) the signal power is split among various nodes, and hence this type of network cannot be used in long distance communication. Mostly broadcast-and-select optical networks are being used in high-speed local area networks and metropolitan area networks.

1.2.3.2 Wavelength-routed optical networks

Wavelength-routed optical network [6, 7, 17] is being designed to overcome the problems of broadcast-and-select network. Wavelength-routed optical network has the potential to solve the problems, mainly (i) lack of wavelength reuse, (ii) power splitting loss and (iii) scalability of wide-area network. A wavelength-routed network consists of routing nodes which are interconnected by fiber links. Each node is equipped with a set of transmitters and receivers for sending and receiving data. In a wavelength-routed optical network, end users communicate with one another via all-optical WDM channels, which are referred to as light-paths [6, 7]. Although use of wavelength converters in an optical network may increase the number of established lightpaths, however they still remain very expensive. Furthermore, uses of wavelength converters introduce an extra traffic delay in the network. Therefore, most of the research in the WDM based optical network focuses mainly on without wavelength conversion. In the absence of wavelength converters, the same wavelength must be used on all hops in the end-to-end path of a connection. This property is known as the wavelength continuity constraint [6, 7]. Figure 1.7 shows the establishment of lightpaths between source-destination pairs on different wavelengths in an example wavelength routed optical network. For the same network, the established lightpaths between source-destination pairs are shown in Table 1.2. In the figure below, each lightpath uses the same wavelength on all hops in the end-to-end path due to wavelength continuity constraint property. The established lightpaths between source-destination pairs A-C and B-F use different wavelengths λ_1 and λ_2, because they use the common fiber link 6-7. This property is known as the distinct channel constraint [6, 7]. The established lightpaths between source-destination pairs H-G and D-E use the same wavelength λ_1 which is already used by the lightpath A-C due to a wavelength reuse characteristic. Given a set of connection requests to be served by the WDM system, the problem of the establishment of lightpaths for each connection request by selecting an optimal route and assigning a required wavelength is known as the routing and wavelength assignment (RWA) problem [6, 7].

A WDM based wavelength-routed optical network [6] has mainly three layers, namely, (i) physical layer, (ii) optical layer and (iii) client layer. Figure 1.8 shows all the possible layers in a WDM based wavelength-routed optical network which are discussed below.

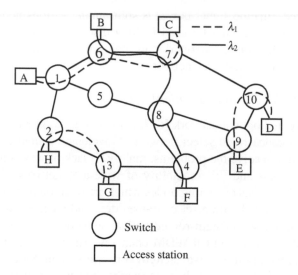

Figure 1.7: A wavelength-routed optical network.

Table 1.2: Summaries of established lightpaths.

S-D pair	Used wavelength	Lightpath
A-C	λ_1	A-1-6-7-C
B-F	λ_2	B-6-7-8-4-F
H-G	λ_1	H-2-3-G
D-E	λ_1	D-10-9-E

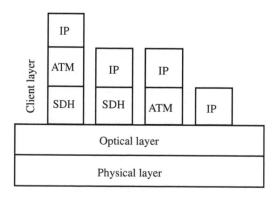

Figure 1.8: Layers of a WDM based wavelength-routed optical network.

(a) **Physical layer**: Physical layer is the lowest layer of an optical network. It is designed to meet the traffic demand, utilize the network resources efficiently and provide quality of service to the end-users.

(b) **Optical layer**: Optical layer is the middle layer, between the lower physical and upper client layers. The optical layer provides lightpaths to the client layers. These lightpaths are the physical links between the client layer network elements. This layer also provides the client independent or protocol transparent circuit-switched service to a variety of clients. Therefore, the optical layer can support a variety of clients simultaneously, for an example, some lightpaths may carry ATM cells, whereas others may carry SONET data or IP packets/datagrams. A WDM based optical network with an optical layer can be configured in such a way that if any failure occurs, the signal can be transmitted using alternate paths automatically. Thus, the reliability of this type of network is higher compared to a traditional network. An optical layer can be further decomposed into three sub layers: (i) optical channel layer, (ii) optical multiplex section layer and (iii) optical transmission section layer. The functionality of an optical channel layer is to provide end-to-end networking of optical channels or lightpaths for transparently conveying the client data. Optical multiplex section layers aggregates low speed multi wavelength optical signals. An optical transmission section layer is concerned with the transmission of optical signals on different kinds of optical media such as single-mode and multi-mode transmission.

(c) **Client layer**: The most common protocols of client layer are SONET/SDH, Ethernet and ATM, which are being used to communicate with end-users. Detailed descriptions of these protocols can be found in [19].

1.3 Limitation of DWDM-based optical networks

In this modern era, everything is becoming dependent upon technology. As a result, the number of traffic in the network grows exponentially. In this context, an optical network is the potential solution to fulfill the exponential traffic demands. The high capacity of DWDM-based optical networks [6,7] is assisted by the use of upper layers to aggregate low-rate traffic flows into lightpaths in the mechanism of traffic grooming [20–24]. DWDM-based systems with up to 40 Gbps capacity per channel have been deployed in backbone networks, while 100 Gbps interfaces are now commercially available, and 100 Gbps deployment is expected soon. TeleGeography [25] expects that international bandwidth demands will be approximately 1,103.3 Tbps in 2020. Therefore, optical networks will be required to support Tb/s class transmission in the near future [26, 27]. Unfortunately, DWDM-based optical transmission technology has an inadequate scaling performance to meet growing traffic demands as it suffers from the electrical bandwidth bottleneck limitation, and the physical impairments become more serious as the transmission speed increases [8]. Moreover, the traffic behavior is changing rapidly and the increasing mobility of traffic sources makes grooming more complex. Therefore, researchers are now focusing on new technologies for high-speed optical networks.

To meet the needs of the future Internet, optical transmission and networking technologies are moving towards the goals of greater efficiency, flexibility, and scalability. Recently, elastic optical networks (EONs) [28–33] have been shown to be promising candidates for future high-speed optical communication, which will be discussed in later chapters.

Exercises

1. What are the advantages of an optic fiber-based communication system over a copper wire-based communication system?

2. What are differences among WDM, FDM and TDM technologies?

3. Discuss the schematic view of a point-to-point DWDM system.

4. Estimate the wavelength spacing when the center wavelength, frequency spacing, and speed of the light in free space are 1520 nm, 80 GHz, and 3×10^8 meters per second, respectively.

5. Consider three regions, which are S band (1460-1530 nm), C band (1530-1565 nm), and L band (1565-1625 nm). Calculate the total number of wavelengths that can be used for each region, when frequency spacing is considered 100 GHz.

6. What are the advantages of synchronous digital hierarchy (SDH) technology over plesiochronous digital hierarchy (PDH) technology?

7. Which functional layer in the SONET is responsible for handling the following functions?

   i A SONET path fails, and the traffic must be switched over to another path.

   ii Multiple SONET streams are to be multiplexed onto a higher-speed stream and transmitted over a SONET link.

   iii A fiber fails, and SONET line terminals at the end of the link reroute all the traffic on the failed fiber onto another fiber.

iv   The error rate on a SONET link between regenerators is to be monitored.

v   The connectivity of an STS-1 stream through a network needs to be verified.

8. Mention various types of physical layer impairments in optical fiber communication?

9. Consider an optical fiber link, where the input power is 0.2 mW and the required minimum power at the receiver is 0.05 mW to establish lightpaths. The loss for the fiber is 0.03 dB/km. How much distance can be transmitted by the signal without considering any amplifier?

10. What is a lightpath? Explain its significance.

11. Suppose that you want to design an optical network. Mention the scenarios in which you would prefer to consider passive-star-based optical networks over wavelength-routed based optical networks and vice verse.

12. In a WDM based network, if two lightpath requests want to use the same wavelength, a conflict may occur. Describe two methods for resolving this conflict.

13. Consider a simple wavelength-routed optical WDM network shown in Fig. 1.7. Considering Fig. 1.7, explain wavelength continuity constraint, distinct channel constraint, and wavelength reuse property.

14. Consider a network mentioned in Fig. 1.9. Assume that each link has two wavelengths. Initially, all wavelengths for each link are available. Consider 15 lightpath requests, which are AB, AC AD, AE, AF, BC, BD, BE, BF, CD, EC, FC, ED, FD, and FE, arrive in the network sequentially. Perform lightpath establishment and estimate the number of blocked requests under the following conditions. If requests are blocked, identify them.

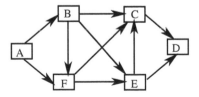

Figure 1.9: Network topology.

    i Consider the minimum hop routing and the first fit wavelength assignment policy, which is described in Chapter 4.

   ii No spectrum conversion is allowed.

  iii No lightpath is tore down after the establishment.

15. If we allow the alternate path routing for lightpath requests AC, AD, and AE on Exercise 14, does it affect the performance on blocking? Provide the analysis.

16. What are the limitations of DWDM based optical networks?

Chapter 2

Elastic Optical Networks

Elastic optical networks (EONs) have been introduced to address the issues of the existing scalable wavelength-division multiplexing (WDM)-based optical networks by providing spectrum-efficient and scalable transport of 100 Gb/s services and beyond. This is achieved through the introduction of flexible granular grooming in the optical frequency domain. An EON has the potential to allocate spectrum to lightpaths according to the bandwidth requirements of clients. The spectrum is divided into narrow slots and optical connections are allocated a different numbers of slots. As a result, network utilization efficiency is greatly improved compared to the WDM-based optical networks. This chapter presents the motivation of the EON, introduces its unique concepts and its enabling technology.

2.1 Motivation behind EONs

The WDM-based optical network divides the spectrum into separate channels. The spacing between adjoining channels is either 50 GHz or 100 GHz, which is specified by the international telecommunication union (ITU)-T standards as shown in Fig. 2.1. The frequency spacing between two adjacent channels is relatively large. If the channels carry only a low bandwidth, and no traffic can be transmitted in the large unused frequency gap, and a large portion of the spectrum will be wasted, which is reflected in Fig. 2.2. To overcome the limitations of traditional optical networks, Jinno et al. [26, 28, 34, 35] presented a spectrum efficient EON based on orthogonal frequency-division multiplexing (OFDM) technology [9, 32]. OFDM is a special class of the multi-carrier modulation (MCM) scheme that transmits a high-speed data stream by dividing it into a number of orthogonal channels, referred to as subcarriers, each carrying a relatively-low data rate [36]. Compared to WDM systems, where a fixed channel spacing between the wavelengths is usually needed to eliminate crosstalk, EONs with OFDM technology allow the spectrum of individual subcarriers to overlap because of its orthogonality, as depicted in Fig. 2.3, which increases the transmission spectral efficiency.

Figure 2.1: ITU-T grid.

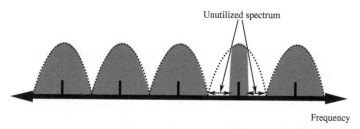

Figure 2.2: Spectrum allocation in WDM based optical networks.

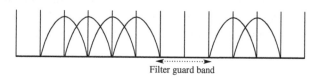

Figure 2.3: Overlapping subcarriers caused by OFDM technology.

2.2   Relationship between OFDM technology and EONs

OFDM [37, 38] splits the data stream into several sub-streams, which are sent in a parallel manner on several subcarriers. Each subcarrier can be modulated based on bit rate requirement and transmission reach [29, 39].

Fig. 2.3 shows the amplitude versus frequency of four different subcarriers. The peak amplitude of any subcarrier's spectrum coincides with the zero point of other subcarriers' spectra, which means that when a subcarrier is sampled at its own peak, all other subcarriers cross at zero point. Thus, they do not interfere each other and no guard band is required to separate them. Due to the orthogonality, OFDM achieves spectrum efficiency.

Keeping the bandwidth requirement constant (the same data rate), a less robust modulation format, such as quadrature phase shift keying (QPSK), carries twice the number of bits per symbol than a more robust modulation format, such as binary phase shift keying (BPSK), which means the baud rate for QPSK is the half of BPSK. When the baud rate of each sub-stream is reduced, we save spectral bandwidth, and hence spectrum efficiency is increased. Therefore, if QPSK is used instead of BPSK, spectrum utilization is enhanced.

The optical signal to noise ratio (OSNR) degradation using QPSK is larger than that of the BPSK. Therefore, the transmission reach using QPSK is shorter than BPSK.

From the above discussion, it can be summarized that OFDM allows us to choose a suitable modulation format considering users' bit rate requirements and transmission reach, which are the key characteristics of EONs.

2.3   Unique concepts of EONs

To develop a better understanding of an EON and its unique properties, let us review several concepts that are not typically part of WDM-based fixed optical networks [30]. The unique characteristics of an EON are bandwidth segmentation, bandwidth aggregation, efficient accommodation of multiple data rates, elastic variation of allocated resources, reach-adaptable line rate, etc. These are discussed in more detail below.

In the fixed WDM-based optical network, there is typically one way to implement a given demand, where bit rate, optical reach, and spectrum are fixed. As a result, the demand can occupy (i) more than one wavelength (in Fig. 2.4(a) to implement a 300 Gb/s demand, multiple wavelengths are required) or (ii) less than a full wavelength, and hence wasting capacity, such as demand C in Fig. 2.4(c). Whereas, EONs allow network operators multiple choices to implement a demand; the channels to implement a 300 Gb/s demand can be grouped tightly into a super-channel and transported as one entity, as shown in Fig. 2.4(b). The possible choices are (i) a given demand can be assigned a modulation format that gives sufficient performance to reach the required distance, while minimizing the spectral bandwidth occupied by the optical path; see demands D and E in Fig. 2.4(d). (ii) Today the ratio between the amount of forward error correction (FEC) and payload is fixed, but it could be made adaptive in EONs to enable greater distances to be reached when the required bandwidth is lower [40]. (iii) Whenever a connection passes through an ROADM, the ROADM acts as a filter that reduces the optical bandwidth for the channel. When this happens over and over, the resulting bandwidth may be too narrow, affecting the quality of the signal and limiting the reach.

The WDM-based optical network requires full allocation of wavelength capacity to an optical path between an end-node pair. However, EONs provide a

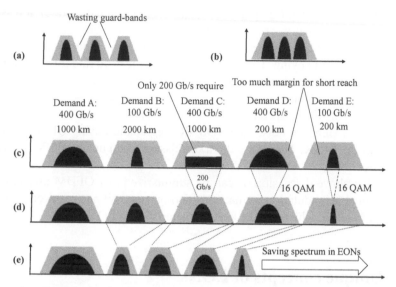

Figure 2.4: Comparison of WDM-based optical networks and elastic optical networks (a) fixed grid requires strict guard-bands between optical paths to implement a 300 Gb/s demand (b) the channels for the demand can be grouped tightly into a super-channel and transported as one entity (c) five demands and their spectrum needs on a 100 GHz fixed grid, assuming quadrature phase shift keying (QPSK) modulation (d) the same demands, with adaptive modulation optimized for the required bit rate and reach (e) the same demands, with additional flexible spectrum.

spectrum efficient bandwidth segmentation (sometimes called sub wavelength) mechanism that provides fractional bandwidth connectivity service. If only partial bandwidth is required, EON can allocate just enough optical bandwidth to accommodate the client traffic, as shown in Fig. 2.5, where a 40 Gb/s optical bandwidth is segmented into three sub wavelengths, such as—5 Gb/s, 15 Gb/s, and 20 Gb/s. At the same time, every node on the route of the optical path allocates a cross-connection with the appropriate spectrum bandwidth to create an appropriate-sized end-to-end optical path. The efficient use of network resources will allow the cost-effective provisioning of fractional bandwidth service.

EONs combine multiple physical ports/links in a switch/router into a single logical port/link to enable incremental growth of link speed as the traffic demand increases beyond the limits of any one single port/link. The EON enables the bandwidth aggregation feature and so can create a super-wavelength optical path contiguously combined in the optical domain, thus ensuring high utilization of spectral resources. This unique feature is depicted in Fig. 2.5, where three 40 Gb/s optical bandwidths are multiplexed with optical OFDM, to provide a super-channel of 120 Gb/s. A super-channel contains multiple very closely spaced

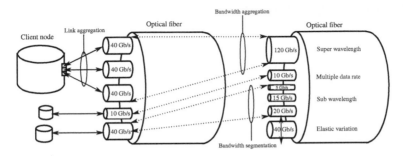

Figure 2.5: Unique characteristics, namely — bandwidth segmentation, bandwidth aggregation, accommodation of multiple data rates, and elastic variation of allocated resources, of elastic optical networks.

channels, which traverse the network as a single entity, but can be demultiplexed at the receiver.

As shown in Fig. 2.5, the EON has the ability to provide the spectrally-efficient direct accommodation of mixed data bit rates in the optical domain due to its flexible spectrum assignment.

It allows network virtualization, which produces virtual networks, consisting of network resources including node and link functionalities. As WDM-based optical networks do not support flexibility and they are strongly integrated with the underlying physical network resources, such as wavelength, it is difficult to utilize the full advantage of network virtualization. However, an EON has the ability to support network virtualization itself as the spectrum resources are segmented and aggregated in order to create a sub-wavelength and super-wavelength channel [28].

As the EON supports a distance-adaptive modulation format, spectrum resources of the fiber are effectively utilized compared to the WDM-based optical networks. In distance a adaptive modulation, the maximum number of bits per symbol, subject to the given transmission characteristic, for an example QoT degradation, is selected for each lightpath. This adaptation is made possible if an efficient format, such as 16-quadrature amplitude modulation (QAM), is selected for shorter distance lightpaths and a more robust one, such as QPSK, is selected for longer distance lightpaths. The necessary number of spectrum slots is then assigned to each lightpath to carry the requested bandwidth.

The EON has the ability to support reach-adaptable line rate [41], as well as dynamic bandwidth expansion and contraction, by altering the number of sub-carriers and modulation formats. It supports energy-efficient operations in order to save power consumption by turning off some of the OFDM subcarriers while traffic is slack.

2.4 Enabling technology for EONs

Advances in optical transmission techniques and devices have favored the emergence of EONs [32, 42–44]. The introduction of advanced modulation formats and wavelength cross-connects (WXCs) enable carrying the growing traffic volume over long-haul distances without optical-electrical-optical (OEO) conversion [45]. The paths with bandwidths determined by the volume of client traffic are allocated through rate-flexible transponders from the transmitter and sent through bandwidth-variable (BV) wavelength cross-connects (WXCs) to the receiver [42].

2.4.1 Spectrally efficient superchannel

There are two common schemes used in rate-flexible transponders to achieve a spectrum efficient modulation for super-channel transmitter [46, 47]. The first scheme is based on OFDM. A frequency-locked multicarrier generator is utilized to generate equally spaced subcarriers. The generated subcarriers are first separated by a wavelength-division demultiplexer (DMUX), then individually modulated with parallel modulators, and finally coupled to generate a spectrally overlapped superchannel. The second scheme is based on WDM of subchannels which have an almost rectangular spectrum with a bandwidth close to the Nyquist limit for intersymbol-interference-free transmission [48]. The subchannels are aligned with the frequency spacing close to the baud rate, which is the Nyquist limit, while avoiding inter-subchannel spectral overlap. This scheme is referred to as Nyquist-WDM [46, 49]. In Nyquist WDM [49], subcarriers are spectrally shaped in order to occupy finer granularity, which corresponds to the baud-rate. These narrow subcarriers are multiplexed at the transmitter with spacing close or equal to the baud-rate, with limited interference and form a super wavelength.

2.4.2 Optical transponders

Three models of transponder can be found [50], namely mixed-line-rate (MLR) model, multi-flow (MF) model, and bandwidth-variable (BV) model. The MLR model employs a few types of transponders, each with a different bit rate, for example 40, 100 and 400 Gb/s transponders to suit a wide range of traffic demands. The MF model uses a MF transponder with several sub-transceivers, that can then be allocated to different demands, each of which has a fixed bit-rate capacity. The BV model supports all types of traffic demands with a single BV transponder, which assigns the fewest possible spectral resources to support traffic demands with a 400 Gb/s maximum bit-rate. As shown in the study presented in [50], the BV model offers better spectrum efficiency and the lowest port consumption rate. Also it is more suited for an energy reduction purpose owing to the reduced active resource due to the use of sub-transceivers. A next generation sliceable bandwidth-variable optical transponder (SBVT) has been investigated in [51]. The authors provided the design architecture and considered several transmission techniques to build the transponder.

Exercises

1. What is the relation between bit rate and baud rate?

2. What are the rationales behind EONs?

3. How does offer OFDM technology better resource utilization than WDM technology?

4. How are sub and super wavelength channels constructed in EONs?

5. What is the relationship between OFDM technology and distance-adaptive modulation?

6. What are the main technical issues of OFDM?

7. Discuss unique properties of EONs over WDM based optical networks.

8. What are the roles of bandwidth-variable optical transponder to design EONs?

9. Consider the C band (1530-1565 nm). Calculate the total number of spectrum slots that can be used, when the frequency spacing is considered 12.5 GHz.

10. Consider a network mentioned in Fig. 2.6. Estimate the number of required slots for each lightpath request, which are AB, AC AD, AE, AF, BC, BD, BE, BF, CD, EC, FC, ED, FD, and FE, under the following conditions.

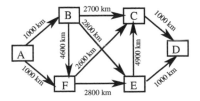

Figure 2.6: Network topology.

i   The routing of each lightpath requests is considered according to the shortest path routing.

ii   The bit rate requirement of each lightpath is 100 Gbps.

iii   Spectrum slot granularity is 12.5 Gbps.

iv   BPSK, QPSK, 8QAM, and 16QAM are used for the transmission reach of > 4500 km, 3500 to 4500 km, 1500 to < 3500 km, and < 1500 km, respectively.

v $m = 1$, $m = 2$, $m = 3$, and $m = 4$ are considered for BPSK, QPSK, 8QAM, and 16QAM, respectively; m represents the modulation level.

Chapter 3

Network and Node Architecture for Elastic Optical Networks

To fulfill our ever-increasing bandwidth demands, the elastic optical network (EON) is indispensable. The performance of the EON depends on its network and node architectures. This chapter presents the architecture of the EON and its operation principle. To complete the discussion of network architecture, this chapter focuses on the different node architecture, and compares their performance in terms of scalability and flexibility.

3.1 Elastic optical network architecture

Figure 3.1 shows the typical architecture of the EON, which mainly consists of Bandwidth-variable transponders (BVTs) and bandwidth-variable cross-connects (BV-WXCs). These basic components and their working principles are explained in the following subsections.

3.1.1 Bandwidth-variable transponder

BVTs [32, 52–55] are used to tune the bandwidth by adjusting the transmission bit rate or modulation format. BVTs support high-speed transmission using spectrally efficient modulation formats, e.g., 16-quadrature amplitude modulation (QAM), with 64-QAM used for shorter distance lightpaths. Longer distance lightpaths are supported by using more robust but less efficient modulation formats, e.g., quadrature phase-shift keying (QPSK) or binary phase-shift keying

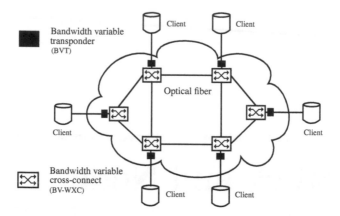

Figure 3.1: Architecture of elastic optical network.

(BPSK). Therefore, BVTs are able to trade spectral efficiency off against transmission reach.

However, when a high-speed BVT is operated at lower than its maximum rate due to required reach or impairments in the optical path, part of the BVT capacity is wasted. In order to address this issue, sliceable bandwidth-variable transponders (SBVTs) [46, 56–59] have been presented that offer improved flexibility; they are seen as promising transponder technology. An SBVT has the capability to allocate its capacity into one or several optical flows that are transmitted to one or several destinations. Therefore, when an SBVT is used to generate a low bit rate channel, its idle capacity can be exploited for transmitting other independent data flows. An SBVT generates multiple optical flows that can be flexibly associated with the traffic coming from the upper layers according to traffic requirements. Therefore, optical flows can be aggregated or can be sliced based on the traffic needs. Figure 3.2 distinguishes BVT and SBVT functionalities.

The SBVT architecture [46, 56] was introduced in order to support sliceability, multiple bit rates, multiple modulation formats, and adaptive code rates. Figure 3.3 shows the architecture of an SBVT; it mainly consists of a source of N equally spaced subcarriers, a module for electronic processing, an electronic switch, a set of N photonic integrated circuits (PICs), and an optical multiplexer. In this architecture, the N subcarriers are generated by a single multi wavelength source. However, such a source may be replaced by N lasers, one per subcarrier. Each client is processed in the electronic domain (e.g., for filtering) and then is routed by the switching matrix to a specific PIC. The generated carriers are equally spaced according to the spectral requirements and the transmission technique adopted. Generated subcarriers are selected at the multi wavelength source, and they are routed to the appropriate PICs. Each PIC is utilized as a single-carrier transponder that generates different modulated signals, such as 16-QAM

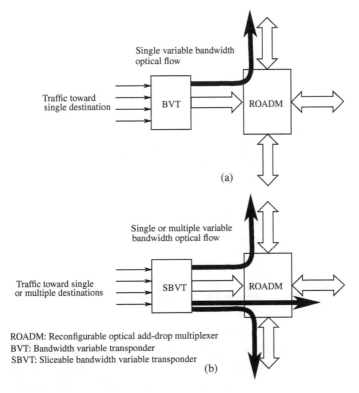

Single variable bandwidth optical flow

Traffic toward single destination

BVT

ROADM

(a)

Single or multiple variable bandwidth optical flow

Traffic toward single or multiple destinations

SBVT

ROADM

ROADM: Reconfigurable optical add-drop multiplexer
BVT: Bandwidth variable transponder
SBVT: Sliceable bandwidth variable transponder

(b)

Figure 3.2: (a) Functionalities of (a) BVT, and (b) SBVT.

and QPSK, in order to support multiple modulation formats. Finally, subcarriers are aggregated by the optical multiplexer in order to form a super channel. Sometimes, subcarriers may be sliced and directed to specific output ports according to the traffic needs. A detailed description of PIC generation of different modulated signals is given in [56].

3.1.2 Bandwidth-variable cross-connect

The BV-WXC [28, 60, 61] is used to allocate an appropriate-sized cross-connection with the corresponding spectrum bandwidth to support an elastic optical lightpath. Therefore, a BV-WXC needs to configure its switching window in a flexible manner according to the spectral width of the incoming optical signal.

Figure 3.4 shows an implementation example of a BV-WXC, where bandwidth-variable spectrum selective switches (BV-SSSs) in the broadcast-and-select configuration are used to provide add-drop functionality for local signals as well as a groomed signal, and routing functionality for transit signals. Typically, a BV-SSS performs wavelength demultiplexing/multiplexing and optical switching functions using integrated spatial optics. The light from an input fiber is di-

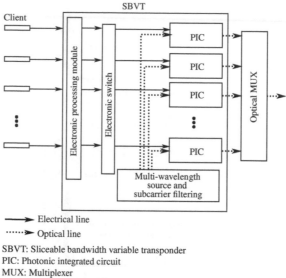

Electrical line
Optical line

SBVT: Sliceable bandwidth variable transponder
PIC: Photonic integrated circuit
MUX: Multiplexer

Figure 3.3: Architecture of SBVT.

vided into its constituent spectral components using a dispersive element. The spatially-separated constituent spectra are focused on a one-dimensional mirror array and redirected to the desired output fiber. Liquid crystal on Silicon (LCoS) or Micro-Electro Mechanical System (MEMS)-based BV-SSSs can be employed as switching elements to realize an optical cross-connect with flexible bandwidth and center frequency. As the LCoS is deployed according to phased array beam steering, which utilizes a large number of pixels, LCoS-based BV-SSSs can easily provide variable optical bandwidth functionality. A detailed description of a BV-WSS employing LCoS technology can be found in [9, 62]. Similarly, details of an MEMS-based BV-SSS can be found in [9, 63].

3.2 Node architectures

This section discusses various node architecture [64–66], which are the building blocks of spectrum efficient EONs.

3.2.1 *Broadcast-and-select*

The broadcast-and-select architecture has been used to determine the elastic optical node architecture that uses spectrum selective switches [65]. Figure 3.5 shows the node architecture of broadcast-and-select, which is implemented using splitters at the input ports. Splitters are used to generate copies of the incoming signals that are subsequently filtered by spectrum selective switches in order to select the required signals at the receiver side. The add/drop network may implement colorless, direction-less, and contention-less elastic add/drop functionality,

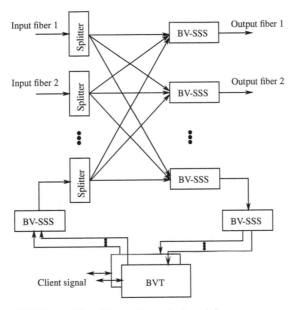

BV-SSS: Bandwidth-variable spectrum selective switch
BV-T: Bandwidth-variable transponder

Figure 3.4: Architecture of BV-WXC.

thus allowing the addition of one or more wavelength channels to an existing multi-wavelength signal automatically. It can also drop (remove) one or more channels from the passing signals to another network path dynamically. The main drawbacks of the broadcast-and-select node architecture are as follows - (i) it requires synchronization and rapid tuning, (ii) it cannot support wavelength reuse and hence a large number of wavelength channels is required, (iii) the signal power is split among various nodes, so this type of node cannot be used for long distance communication. The broadcast and select architecture is mostly being used in high-speed local area networks and metropolitan area networks. It must be noted that the broadcast-and-select architecture struggles to support additional functionality to cope with dynamic requirements, e.g., spectrum defragmentation [67–69].

3.2.2 Spectrum routing

The spectrum routing node architecture is being designed to overcome the problems with the broadcast-and-select node architecture. It is basically implemented with arrayed waveband gratings [9] and optical switches as shown in Fig. 3.6. In spectrum routing, both switching and filtering functionalities are controlled by the spectrum selective switches. The basic advantage of this architecture, com-

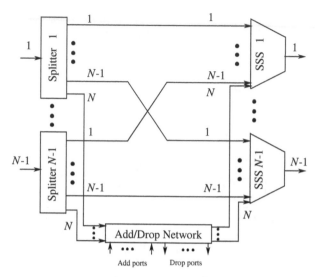

SSS: Spectrum selective switch

Figure 3.5: Node architecture of broadcast-and-select.

pared to the broadcast-and-select architecture, is that the power loss is not dependent on the number of degrees. However, it requires additional spectrum selective switches at the input fibers, which makes it more expensive to realize. Furthermore, the additional functionality needed to cope with dynamic requirements, e.g., spectrum defragmentation [67–69], is still difficult to implement in this architecture.

3.2.3 Switch and select with dynamic functionality

We have already observed that the broadcast-and-select architecture and spectrum routing architecture are unable to support dynamic requirements, such as, spectrum defragmentation, time multiplexing, regeneration, etc. To overcome these limitations, the switch and select architecture with dynamic functionality has been introduced. In this architecture, an optical switch is used to direct copies of the input to a specific spectrum selective switch or to a module (f) that provides additional functionalities, such as—defragmentation, time multiplexing, and regeneration. The outputs of the modules connect to spectrum selective switches, where the required signals are filtered for delivery to the corresponding output fiber. Figure 3.7 shows the node architecture of the EON with the dynamic functionalities that support dynamic requirements, namely—spectrum defragmentation, time multiplexing, and regeneration. These dynamic functionalities come at the price of additional large port count optical switches and larger

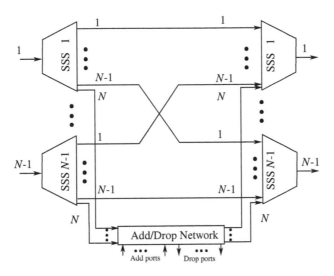

SSS: Spectrum selective switch

Figure 3.6: Node architecture of spectrum routing.

spectrum selective switch port counts. The number of ports is dedicated to provide a specific functionality, and hence the number of modules may be calculated from the expected demand.

3.2.4 Architecture on demand

The architecture on demand (AoD) [70] consists of an optical backplane that is implemented with a large port-count optical switch connected to several processing modules, namely—spectrum selective switch, fast switch, erbium-doped fiber amplifier (EDFA), spectrum defragmenter, splitter, etc. The inputs and outputs of the node are connected via the optical backplane as shown in Fig. 3.8. The different arrangements of inputs, modules, and outputs are realized by setting appropriate cross connections in the optical backplane. Therefore, it provides greater flexibility than the architecture as explained above. This is mainly due to the non-mandatory nature of the components (such as—spectrum selective switch, power splitters and other functional modules) unlike static architectures, but they can be interconnected together in an arbitrary manner. The number of spectrum selective switches and other processing devices is not fixed but can be determined based on the specific demand for those functionalities. Thus, savings in the number of devices can balance the additional cost of the optical backplane, and hence this type of architecture provides a cost-efficient solution. Furthermore, AoD provides considerable gains in terms of scalability and resiliency compared to other static architectures.

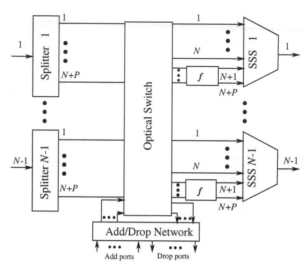

SSS: Spectrum selective switch

Figure 3.7: Node architecture of switch and select with dynamic functionality.

3.2.5 Comparing node architectures

Table 3.1 summarizes the above discussed node architectures in terms of total power loss, port count of switch/backplane, routing flexibility, port count of spectrum selective switches, defragmentation capability, time multiplexing, and regeneration capability. The calculation of total power loss [65] is determined by the type of node architecture implemented. In case of AoD, total power loss depends on the architecture implemented and the number of cross connections used in the optical backplane. The total power loss in the switch and select with dynamic functionality architecture depends on the spectrum selective switches, backplane, and modules used. However, the total power losses of the broadcast-and-select architecture and spectrum routing architecture mainly depend on the spectrum selective switches.

Port count of switch/backplane [65] varies the networking cost. The switch and select node and AoD node architectures need optical switches. However, the number of SSSs and other processing devices may be tailored to suit the specific demand. Therefore, as savings the number of devices can offset the additional cost of the optical backplane, it provides an overall cost-effective solution. The number of SSS ports required by an AoD node is not strictly related to the node degree. This is because several small-port-count SSSs may be connected together in order to increase the number of available ports.

Table 3.1: Comparison of different node architectures for the EON.

	Broadcast-and-select	Spectrum routing	Switch and select with dynamic functionality	Architecture-on-demand
Power loss [in dB] [65]	$3\lceil\log_2(N)\rceil + L_{SSS}$	$2L_{SSS}$	$3\lceil\log_2(N + P)\rceil + L_{SSS} + L_{Switch} + L_M$	Switch and select architecture $+ (m+1)L_{Switch}$ for path with m modules
Switch/backplane port count [65]	Not required	Not required	$N(N+P)$	$2(N-1) + N(N+P)$
Routing flexibility	No	No	Medium	High
SSS port count	N	N	$N+P$	P
Defragmentation	No	No	Yes	Yes
Time multiplexing	No	No	Yes	Yes
Regeneration	No	No	Yes	Yes

SSS: Spectrum selective switch, L_{SSS}: SSS loss, L_{Switch}: Switch/backplane loss, L_M: Module loss.

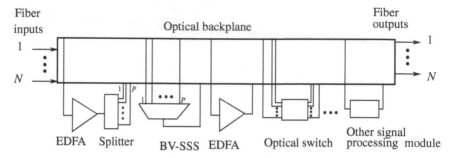

EDFA: Erbium-doped fiber amplifier
BV-SSS: Bandwidth-variable spectrum selective switch

Figure 3.8: Node architecture on demand with N input/outputs, and signal processing modules.

Routing flexibility is the capability of the system to carry signals from source to destination along different routes. This type of flexibility is required when strengthening system resilience to failures along working paths; signals may be directed to their backup paths. Time multiplexing is used to transmit and receive independent signals over a common signal path by synchronized switches at each end of the transmission line. As a result, each signal appears on the line only a fraction of the time in an alternating pattern. On the other hand, all-optical 3R (Re-amplification, Re-shaping, and Re-timing) signal regeneration is needed to avoid the accumulation of noise, crosstalk and non-linear distortion, and to ensure a good signal quality for transmission over any path in an optical network. Spectral defragmentation is a technique to reconfigure the network so that the spectral fragments can be consolidated into contiguous blocks.

Exercises

1. Compare SBVT and BVT.

2. What is the main advantage of spectrum routing node architecture over broadcast-and-select node architecture.

3. What is the main advantage of switch and select with dynamic functionality node architecture over spectrum routing node architecture?

4. Describe the properties that a node should exhibit in order to be an appropriate candidate for a flexible node architecture.

5. Describe the comparison of different node architectures.

6. Why does AoD consider one of the suitable node architectures for EONs?

Chapter 4

Routing and Spectrum Allocation for Elastic Optical Networks

Routing and spectrum allocation (RSA) in elastic optical networks (EONs) is considered one of the key functionalities due to its information transparency and spectrum reuse characteristics. RSA is used to (i) find the appropriate route for a source and destination pair, and (ii) allocate suitable spectrum slots to the requested lightpath. The RSA problem involves two basic constraints, which are continuity and contiguity constraints. This chapter reviews and classifies RSA approaches, including their pros and cons.

4.1   RSA vs RWA

The RSA problem [71–76] in EONs is equivalent to the routing and wavelength assignment (RWA) problem in wavelength-division multiplexing (WDM)-based optical networks. The problem of establishing lightpaths for each connection request by selecting an appropriate route and assigning the required wavelength is known as the RWA problem [6,7,77]. In WDM-based optical networks without wavelength converters, the same wavelength must be used on all hops in the end-to-end path of a connection. This property is known as the wavelength continuity constraint.

The difference between RSA and RWA is due to the capability of the EON architecture to offer flexible spectrum allocation to meet the requested data rates.

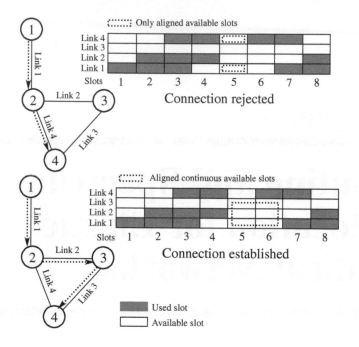

Figure 4.1: Example of continuity and contiguity constraints.

In RSA, a set of contiguous spectrum slots is allocated to a connection instead of the wavelength set by RWA in fixed-grid WDM-based networks. These allocated spectrum slots must be placed near to each other to satisfy the spectrum contiguity constraint. If enough contiguous slots are not available along the desired path, the connection can be broken up into small multiple demands. Each one of these smaller demands would then require a lower number of contiguous subcarrier slots. Furthermore, the continuity of these spectrum slots should be guaranteed in a similar manner as demanded by the wavelength continuity constraint. If a demand requires t units of spectrum, then t contiguous subcarrier slots must be allocated to it (due to the spectrum contiguity constraint), and the same t contiguous slots must be allocated on each link along the route of the demand (due to the spectrum continuity constraint).

The concept of the contiguity and continuity constraints of the spectrum allocation is explained with an example. For this purpose, we consider the network segment shown in Fig. 4.1. We assume a connection request that requires a bitrate equivalent to two slots for RSA from source node 1 to destination node 4. The connection request cannot be established through the shortest route 1-2-4 because the links from 1-2 and 2-4 have two contiguous slots that are not continuous, so the continuity and contiguity constraints are not satisfied. However, the continuity and contiguity constraints are satisfied if the connection uses the route 1-2-3-4, and spectrum slots 5 and 6.

RWA in WDM-based optical networks is an nondeterministic polynomial time (NP)-hard problem, and has been well studied over the last twenty years. The RWA problem is reducible to the RSA problem as the number of wavelengths equals the number of spectrum slots in each fiber link. For any lightpath request, if RWA requires 1 wavelength along the lightpath, it is equivalent to a 1 spectrum slot request along the lightpath in the RSA problem. This reduction is within poly-nominal time. The RWA problem has a solution if and only if the constructed RSA problem has a solution. Therefore, from the above discussion, we can say that the RSA problem is an NP-hard problem [31, 78].

Although RSA is a hard problem, it can simplified by splitting it into two separate subproblems, namely—(i) the routing subproblem, and (ii) the spectrum allocation subproblem. These subproblems are discussed in section 4.2 and section 4.3, respectively.

4.2 Routing

Approaches for solving the routing subproblem in the EON fall into two main groups, namely—(i) routing without elastic characteristics, and (ii) routing with elastic characteristics. In the following, we explain these two routing approaches.

4.2.1 *Routing without elastic characteristics*

This subsection focuses on different routing policies [79–84], namely—(i) fixed routing, (ii) fixed alternate routing, (iii) least congested routing, and (iv) adaptive routing, with no consideration given to the elastic characteristics of optical networks. These routing approaches are mainly intended to discover suitable routes between source-destination pairs. These algorithms are discussed below.

4.2.1.1 *Fixed routing*

In fixed routing (FR) [6, 80], a single fixed route is precomputed for each source-destination pair using some shortest path algorithms, such as Dijkstra's algorithm [85]. When a connection request arrives in the network, this algorithm attempts to establish a lightpath along the predetermined fixed route. It checks whether the required slot is available on each link of the predetermined route or not. If even one link does not have the slot desired, the connection request is blocked. In the situation when more than one required slot is available, a spectrum allocation policy is used to select the best slot.

4.2.1.2   Fixed alternate routing

Fixed alternate routing (FAR) [6,80] is an updated version of the FR algorithm. In FAR, each node in the network maintains a routing table (that contains an ordered list of a number of fixed routes) for all other nodes. These routes are computed off-line. When a connection request with a given source-destination pair arrives, the source node attempts to establish a lightpath through each of the routes from the routing table taken in sequence, until a route with the required slot is found. If no available route with required slot is found among the list of alternate routes, the connection request is blocked. In the situation when more than one required slot is available on the selected route, a spectrum allocation policy is used to choose the best slot. Although the computation complexity of this algorithm is higher than that of FR, it provides comparatively lower blocking probability than the FR algorithm. However, this algorithm may not be able to find all possible routes between a given source-destination pair. Therefore, the performance of the FAR algorithm in terms of blocking probability is not optimum.

4.2.1.3   Least congested routing

Least congested routing (LCR) [6, 80] predetermines a sequence of routes for each source-destination pair similar to FAR. Depending on the arrival time of connection requests, the least-congested routes are selected from among the pre-determined routes. The congestion on a link is measured by the number of slots available on the link. If a link has fewer available slots, it is considered to be more congested. The disadvantage of LCR is its higher computation complexity; its blocking probability is almost the same as that of FAR.

4.2.1.4   Adaptive routing

In adaptive routing (AR) [80, 81], routes between source-destination pairs are chosen dynamically, depending on the network status, such as link-state information. The network status is determined by the set of all connections that are currently active. The most acceptable form of AR is adaptive shortest path routing, which is well suited for use in optical networks. Under this approach, each unused spectrum in the network has a cost of 1 unit, whereas the cost of each used spectrum in the network is taken to be α. When a connection arrives, the shortest path between a source-destination pair is determined. If there are multiple paths with the same distance, one of them is chosen at random. In AR, a connection is considered blocked mainly when there is no route with a required slot between the source-destination pair. Since AR considers all possible routes between source-destination pair, it provides lower blocking probability, but its setup time is comparatively higher than other routing policies. AR requires extensive support from control and management protocols to continuously update the routing tables at the nodes. AR suits centralized implementation rather more than the distributed alternative.

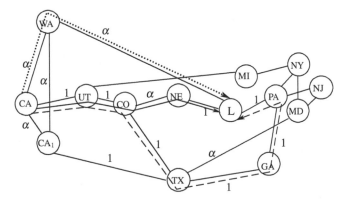

Figure 4.2: Fixed (solid line), alternate (dotted line) and adaptive (dashed line) routes are shown from source city CA to destination city L.

The functionality of the above mentioned routing policies is explained with the help of the sample example network, see Fig. 4.2. It consists of 14 nodes (representing cities) and 21 bi-directional optical links. The fixed shortest route, alternate route, and adaptive route from source city CA to destination city L are shown by the solid, dotted, and dashed lines, respectively. Furthermore, the congested links are denoted as α. If a connection request for a connection from source city CA to destination city L arrives, only AR is able to find a route between CA and L.

4.2.1.5 Comparisons of routing policies

A significant amount of work on the different issues of routing has been reported. Table 4.1 summarizes the major routing policies, and compares their performance in terms of blocking probability, and average setup time [77]. The blocking probability [86, 87] in the network is defined as the ratio of the number of blocked lightpath requests to the number of lightpath requests in the network. The average setup time [23] in the network is defined as total execution time required to establish all the lightpaths in the network to the number of successful lightpaths. We observe that FR has the lowest average setup time and time complexity of all routing policies. However, its blocking probability is the highest. AR provides the best performance in terms of blocking probability, but its time complexity is the highest. FAR offers a trade-off between time complexity and blocking probability.

Table 4.1: Summaries of different routing policies.

Problem	Approach		Reference	Performance analysis		On/Off line
				BP	AST	
Routing	Static	FR	[6,7]	Higher BP than others	Shorter AST than others	Off line
		FAR	[6,7]	Lower BP than FR	Longer AST than FR	Off line
		LCR	[6,7]	Almost same as FAR	Almost same as FAR	Off line
	Dynamic	AR	[6,7]	Lower BP than others	Longer AST than others	On line

4.2.2 Routing with elastic characteristics

An EON has the capability to slice the spectrum into slots with finer granularity than WDM-based networks. Jinno et al. [29] presented, for first time, the method referred to as single slot on the grid approach, see Fig. 4.3. In this approach, the frequency slots are based on the ITU-T fixed grids, where the central frequency is set at 193.1 THz. The width of a frequency slot depends on the transmission system. In this example, one frequency slot is 12.5 GHz. According to the bandwidth demand of a connection request, a group of frequency slots, usually consecutive in the frequency domain, are allocated.

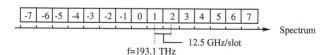

Figure 4.3: Frequency slot approach for elastic optical networks.

In EONs, single path routing via the RSA approach can create the spectrum fragmentation problem and thus inefficiency. The spectrum fragmentation issue is explained in detail in the following chapters. To overcome this problem, multi-path routing [88–93] has been considered for the EON. An example of this routing is shown in Fig. 4.4. We consider a lightpath request $L(F,S,D)$, where F, S, and D are the number of required contiguous slots, source, and destination, respectively. We assume that lightpath requests arrive in the system in serial manner. We refer to the available slots as fragmented slots after the spectrum allocation of lightpath requests $L_1, L_2, \cdots, L_6$. In this context, if lightpath request L_7 arrives at node A for destination node C with demand of four consecutive slots, it is rejected as the required slots are unavailable. However, two lightpaths (i.e., $A-B-C$ and $A-D-C$), each of which will utilize two consecutive slots, can be setup to service L_7. This type of routing, called multi-path routing or sometimes spectrum-split routing, is caused by spectrum fragmentation. Multi-path routing

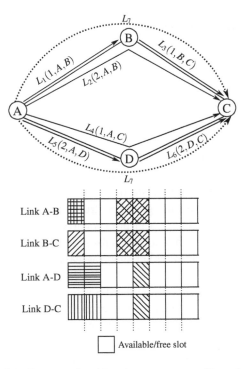

Figure 4.4: Concept of multi-path or spectrum-split routing in EONs.

can be used to handle the spectrum fragments that are very common in dynamic traffic scenarios.

4.3   Spectrum allocation

With the aim of better fitting the bandwidth requirements at each moment, the lightpaths established in a network may dynamically change their allocated spectrum. This capability is defined as elastic spectrum allocation [94, 95] and its implementation in future flexgrid networks is expected to provide better network performance. Spectrum allocation may be performed either after finding a route for a lightpath or in parallel during the route selection process. This section discusses the different spectrum allocation policies. We categorize the spectrum allocation based on spectrum range for connection groups and spectrum slot for individual connection request, as presented below.

4.3.1   Spectrum range allocation for connection groups

Policies used to allocate spectrum range for connection groups can be categorized into three types, namely—(i) fixed spectrum allocation, (ii) semi-elastic spectrum allocation and (iii) elastic spectrum allocation, based on the changes allowed to the resources allocated to lightpaths in terms of central frequency (CF) and spectrum width.

4.3.1.1   Fixed spectrum allocation

In the fixed spectrum allocation (fixed SA) policy [94,95], both CF and assigned spectrum width remain static for ever. At each time period, demands may utilize either whole or only a fraction of the allocated spectrum to convey the bit rate requested for that period. Therefore, this policy does not provide any elasticity. Under this policy, the spectrum allocation of lightpaths is independent of variations of bandwidth requirements. When the bandwidth demand of a connection is lower than the capacity of the assigned spectrum, the connection request is established. In this case, the utilized spectrum for carrying traffic is lower than that of allocated spectrum. This can lead to a sub-optimal use of network capacity. When the bandwidth demand is higher than the capacity of assigned spectrum, some bandwidth is not served.

Figure 4.5 shows the concept of the fixed-SA policy. The bandwidth demand for the lightpath at time T is equal to the capacity of the assigned spectrum. In an underused spectrum condition, the bandwidth demand is lower than the capacity of the assigned spectrum in time T' as shown in Fig. 4.5(a). Similarly, the bandwidth demand is higher than the capacity of the assigned spectrum in time T' under insufficient spectrum condition as shown in Fig. 4.5(b).

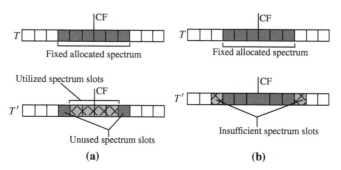

Figure 4.5: Different conditions (a) underused spectrum condition, and (b) insufficient spectrum condition of fixed spectrum allocation policy.

4.3.1.2   Semi-elastic spectrum allocation

In the semi-elastic spectrum allocation (semi-elastic SA) policy [94,95], the CF remains fixed, but the allocated spectrum width can vary in each time interval. The frequency slices are allocated to a lightpath so as to suit the required bandwidth at any time. As a result, the unutilized slots can be used for subsequent connection requests. Therefore, this spectrum allocation policy provides higher flexibility than the fixed SA policy. To explain semi-elastic SA, two scenarios are considered below.

(i) If the required bandwidth demand is reduced, the capacity of the allocated spectrum can also be reduced. The unnecessary spectrum slices at each end of the allocated spectrum can be released and may be used for subsequent connection requests. Figure 4.6(a) represents a spectrum slot reduction condition, where both utilized and allocated spectra of the channel occupy the same number of slices at time T'.

(ii) If the required bandwidth demand is increased, new contiguous spectrum slices can be allocated at both ends of the CF. The capacity of the allocated spectrum can be increased in order to serve the maximum required bandwidth. Figure 4.6(b) represents a spectrum slot expansion condition, where a lightpath increases its required bandwidth from six to eight slots.

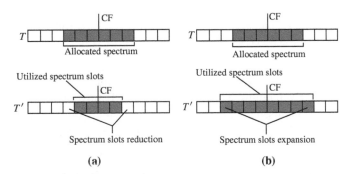

Figure 4.6: Different conditions of the semi-elastic spectrum allocation policy with (a) spectrum slot reduction, and (b) spectrum slot expansion.

4.3.1.3 Elastic spectrum allocation

In the elastic spectrum allocation (elastic SA) policy [94, 95], both assigned CF and the spectrum width can be changed in each time interval. This spectrum allocation policy adds a new degree of freedom to the previous policy. It not only allows the number of slots per lightpath to be varied at any time, but also CF can be changed. Furthermore, we distinguish two conditions that differ in the grade of flexibility of CF movement as follows.

(i) CF movement within a range: As CF movement is limited to a certain range, the spectrum reallocation is restricted to neighboring CFs. Figure 4.7(a) represents a lightpath that varies its requirement at time T and T'. In this figure, both the assigned spectrum width and the CF are varied, but the CF is varied within the range specified.

(ii) Elastic spectrum reallocation: This condition reallocates the spectrum completely, and there is no CF movement limitation. The elastic spectrum allocation policy offers the best spectrum utilization performance among all spectrum allocation policies. Figure 4.7(b) represents a typical elastic spectrum allocation policy, where two lighpaths are reallocated and their allocated spectrum widths are varied.

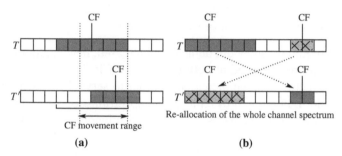

(a)   (b)

Figure 4.7: Different conditions of the elastic spectrum allocation policy with (a) CF movement within a range, and (b) elastic spectrum reallocation.

A comparison of the CF movement approach within a range and the reallocation approach is given below. The first one limits CF movement and so has lower hardware requirements. However, the limitation yields only limited flexibility, while the elastic SA policy with reallocation approach is completely flexible.

4.3.2 Comparison of spectrum range allocation policies for connection groups

Table 4.2 summarizes the different policies available to allocate spectrum range for connection groups in terms of hardware requirement, control plane, complexity of signal procedures, computation complexity, spectrum utilization, and nature of spectrum allocation. We observe that both CF and assigned spectrum width remain static in fixed SA policy. However, the control plane can be configured in order to allocate a fixed channel consisting of a fixed number of slices. The main drawback of the fixed spectrum allocation policy is the sub-optimal use of capacity, which makes this policy less convenient.

Compared to the fixed SA policy, the semi-elastic SA policy requires additional hardware. Moreover, the control plane can be configured in such way that it can allow established lightpaths to be modified. Since the amount of frequency slices are assigned to each lightpath dynamically, extension of the RSVP-TE protocol [96] should be designed so as to notify all BV-WXCs along the path to adjust their filter bandwidth, and modify the number of slices allocated to a path. Furthermore, some hardware is required to increase or decrease the utilized spectrum as needed. Therefore, bandwidth variable transponders and bandwidth

Table 4.2: Summaries of different spectrum range allocation policies for connection groups.

	Fixed SA	Semi-elastic SA	Elastic SA
Hardware Requirement	No extra requirement	BVT and BV-WXC should allow to increase or decrease the number of allocated slices.	Higher requirements for the range of laser tunability
Control Plane	Extension: GMPLS, RSVP-TE, OSPF	Extension in RSVP-TE protocol to modify the number of assigned slices.	Complex algorithms in PCE to prevent conflict (simultaneous access to spectrum resources).
Complexity of signal procedures	Control plane does not need to notify BV-WXCs along the path to adjust filter bandwidth.	RSVP-TE protocol is used to notify all BV-WXCs along the path to adjust their filter bandwidth.	Similar procedure as semi-elastic SA and extra waiting time to adjust filters and lasers
Computation Complexity	Lower time complexity than others	Higher time complexity than Fixed SA, but lower than Elastic SA	Highest time complexity among all
Spectrum Utilization	Lower spectrum utilization than others	Higher spectrum utilization than Fixed SA, but lower than Elastic SA	Highest spectrum utilization among all
On/Off line	Off line	On line	On line

SA: Spectrum allocation, BVT: Bandwidth-variable transponder, BV-WXC: Bandwidth-variable wavelength cross-connect, GMPLS: Generalized multi-protocol label switching, RSVP-TE: Resource reservation protocol - traffic engineering, OSPF: Open shortest path first, PCE: Path computation element

variable switches should work with frequency steps in accordance with the frequency slice width. The semi-elastic SA policy has better performance than of the fixed SA policy in terms of spectrum efficiency at the cost of some extra hardware resources.

In the elastic SA policy, extra hardware and a control plane are required to vary both CF and spectrum width dynamically. As both CF and spectrum width vary dynamically, the elastic SA policy provides best performance in terms of spectrum utilization. However, the computation complexity and extra hardware requirements are high compared to other spectrum allocation policies.

4.3.3 Spectrum slot allocation for individual connection request

The spectrum slot allocation of an individual connection request can be performed using one of the following allocation policies [97].

4.3.3.1 First fit

In the first fit spectrum allocation policy [98, 99], the spectrum slots are indexed and a list of indexes of available and used slots is maintained. This policy always attempts to choose the lowest indexed slot from the list of available slots and allocates it to the lightpath to serve the connection request. When the call is completed, the slot is returned to the list of available slots. By selecting spectrum in this manner, existing connections will be packed into a smaller number of spectrum slots, leaving a larger number of spectrum slots available for future use. Implementing this policy, does not require global information of the network. The first fit spectrum allocation policy is considered to be one of the best spectrum allocation policies due to its lower blocking probability and computation complexity.

4.3.3.2 Random fit

In the random fit policy [6, 98], a list of free or available spectrum slots is maintained. When a connection request arrives in the network, this policy randomly selects a slot from the list of available slots and allocates it to the lightpath used to serve the connection request. After assigning a slot to a lightpath, the list of available slots is updated by deleting the used slot from the free list. When a call is completed, its slot is added to the list of free or available slots. By selecting spectrum in a random manner, it can reduce the possibility of multiple connections choosing the same spectrum which is possible if spectrum allocation is performed in a distributed manner.

4.3.3.3 Last fit

This policy [6, 100] always attempts to choose the highest indexed slot from the list of available slots and allocates it to the lightpath to serve the connection request. When the call is completed, the slot is returned back to the list of available slots.

4.3.3.4 First-last fit

In the first-last fit spectrum allocation policy [99], all spectrum slots of each link can be divided into a number of partitions. The first-last fit spectrum allocation policy always attempts to choose the lowest indexed slots in the odd number partition from the list of available slots. For the even number partitions, it attempts to choose the highest indexed slots from the list of available spectrum slots. Use of first fit and random fit spectrum allocation approaches always attempt to choose the lowest indexed slots for each partition and randomly selects slots, respectively, from the list of available spectrum slots. This may lead to a situation where spectrum slots may be available, but connection requests cannot be established due to unavailability of contiguous aligned slots. The first-last fit allocation policy is expected to give more contiguous aligned available slots than the random fit and first fit allocation policies.

4.3.3.5 Least used

The least used spectrum allocation policy [6,7] allocates a spectrum to a lightpath from a list of available spectrum slots that have been used by the fewest fiber links in the network. If several available spectrum slots share the same minimum usage, the first fit spectrum allocation policy is used to select the best spectrum slot. Selecting spectrum in this manner is an attempt to spread the load evenly across all spectrum slots.

4.3.3.6 Most used

The most used spectrum allocation policy [6, 7] assigns spectrum to a lightpath from a list of available spectrum slots, which have been used by the most fiber links in the network. Similar to the least used spectrum allocation policy, if several available spectrum slots share the same maximum usage, first fit spectrum allocation policy is used to break the tie. Selecting spectrum slots in this way is an attempt to realize maximum spectrum reuse in the network.

4.3.3.7 Exact fit

Starting from the beginning of the frequency channel, the exact fit allocation policy [98] searches for the exact available block in terms of the number of slots requested for the connection. If there is a block that matches the exact size of requested resources, this policy allocates that spectrum. Otherwise, the spectrum is allocated according to the first fit spectrum allocation policy. By selecting spectrum slots in this way, we can reduce the fragmentation problem in optical networks.

To illustrate the functionality of the above mentioned spectrum allocation policies, we use the example shown in Fig. 4.8. If two connection requests arrive that use link 2 and link 3 with one slot demand for establishing lightpaths, the strategies proceed as follows. First fit spectrum allocation policy selects spectrum slot 2 for the first connection request, and slot 3 for the second connection request. First-last fit spectrum allocation policy selects spectrum slot 2 for the first connection request, and slot 12 for the second connection request. Slot 6 and slot 4 have been used three times and two times, respectively, in this example. Therefore, slot 6 and slot 4 are used by most used spectrum allocation policies. As slot 2 and slot 9 have not been used so far, the least used spectrum allocation policy selects these two slots for the two connections requests. Random fit allocation policy selects any two of slot 2, slot 3, slot 4 and slot 12 with equal probability. Exact fit spectrum allocation policy selects spectrum slot 6 for the first connection request, and slot 2 for the second connection request.

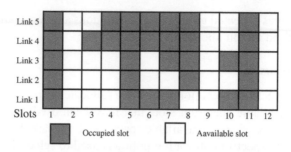

Figure 4.8: Spectrum slot usage pattern for a network segment.

4.3.4 Comparisons of spectrum allocation policies for individual connection requests

A significant amount of published research has addressed different policies to allocate spectrum slots to individual connection requests. Table 4.3 summarizes some major spectrum allocation policies. The least-used and most-used allocation policies have higher time complexity than the other allocation policies. These two spectrum allocation policies also require global information of the network. As random fit, first fit, last fit, exact fit, and first-last fit spectrum allocation policies have lower time complexity, we present the performance of these spectrum allocation policies in terms of blocking probability under differ traffic loads (in Erlang), see Table 4.4. These numerical results were obtained from a simulation study performed on NSFNET. The simulation parameters followed those of [100]. We observe from the numerical results that first-last fit is of lower blocking probability than the other spectrum allocation schemes, as it provides less fragmentation than the other spectrum allocation policies. The blocking probability of first-exact fit is higher than that of first-last fit, but its blocking probability is lower than those of other spectrum allocation policies. First fit and last fit spectrum allocation policies provide almost similar performance. Finally, the blocking probability of random fit is highest among all spectrum allocation policies.

4.3.5 Joint RSA

We have already discussed routing and spectrum allocation in EONs separately in section 4.2 and section 4.3, respectively. However, many researchers have presented joint RSA [101–103] by considering routing and spectrum allocation at the same time. They usually employ a matrix to describe link or path spectral status by considering spectrum continuity and contiguity constraints, and choose the best performance from among all available matrix candidates.

In this direction, Liu et al. [101] presented a layer-based approach to design integrated multicast-capable routing and spectrum assignment (MC-RSA)

Table 4.3: Summaries of different policies to allocate spectrum slots to a single connection request.

Allocation policy	Reference	Applicable network
Least used	[6, 7]	Single/multi-fiber
Most used	[6, 7]	Single/multi-fiber
Random fit	[6, 98]	single/multi-fiber
First fit	[98, 99]	single/multi-fiber
Last fit	[6, 100]	single/multi-fiber
Exact fit	[98]	single/multi-fiber
First-last fit	[99]	single/multi-fiber

Table 4.4: Numerical results of different spectrum allocation policies in RSA in terms of blocking probability.

Allocation policy	Blocking probability				
	20 [Erl]	40 [Erl]	60 [Erl]	80 [Erl]	100 [Erl]
First-last fit	0.005	0.007	0.010	0.017	0.025
First-exact fit	0.006	0.008	0.012	0.020	0.027
First fit	0.006	0.009	0.014	0.021	0.029
Last fit	0.007	0.008	0.015	0.020	0.030
Random fit	0.017	0.022	0.031	0.043	0.055

algorithms for serving multicast requests efficiently and minimizing the blocking probability in the EON. For each multicast request, the presented algorithms decompose the physical topology into several layered auxiliary graphs according to the network spectrum utilization. Then, based on the bandwidth requirement, a proper layer is selected, and a multicast light-tree is calculated for the layer. These procedures realize routing and spectrum assignment for each multicast request in an integrated manner. Similarly, two joint routing and spectrum allocation algorithms [102], namely—(i) fragmentation-aware RSA and (ii) fragmentation-aware RSA with congestion avoidance, have been presented by Yin et al. to alleviate spectral fragmentation in the lightpath provisioning process.

There are some works [31, 104–110] that focus on solving joint RSA with modulation selection. This type of problem is referred to as the routing, modulation, and spectrum allocation (RMSA) problem.

Exercises

1. Why is spectrum contiguity required in EONs?

2. Discuss the pros and cons of different routing policies without considering elastic characteristics.

3. Compare the characteristics of various spectrum allocation policies.

4. Consider a network mentioned in Fig. 4.9. Assume that each link has five spectrum slots. Initially, all slots for each link are available. Consider 15 lightpath requests, AB, AC AD, AE, AF, BC, BD, BE, BF, CD, EC, FC, ED, FD, and FE, arrive in the network sequentially, where each lightpath request is indexed by $i \in \{1, \cdots, 15\}$. Perform spectrum allocation and estimate the number of blocked requests under the following conditions. If requests are blocked, identify them.

    i   Consider the minimum hop routing and the first fit spectrum allocation policy.

    ii Each request requires two contiguous slots for lightpath establishment, and no spectrum conversion is allowed.

    iii No lightpath is torn down after the establishment.

    iv No gardband is considered.

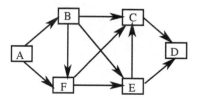

Figure 4.9: Network topology.

5. If we allow the alternate path routing for lightpath requests AC, AD, and AE on Exercise 4, does it affect the performance on blocking? Provide the analysis.

6. If we consider first-last fit on Exercise 4, does it affect the performance on blocking? Provide the analysis. For the first-last fit policy, odd and even indexed lightpath requests are allocated using first fit and last fit, respectively.

7. If we consider spectrum split routing on Exercise 4, does it affect the performance on blocking? Provide the analysis.

8. Why does elastic spectrum allocation provide a lower blocking ratio than fixed and semi-elastic spectrum allocation?

9. Why is first fit spectrum allocation considered the most appropriate spectrum allocation approach? Explain it.

10. Analyze the time complexity for different routing and spectrum allocation policies.

Chapter 5

Different Aspects Related to Routing and Spectrum Allocation

The performance of the elastic optical network (EON) depends not only on its physical resources, like—transponders, physical links, usable spectral width, optical switches, etc., but also how the network is controlled. The objective of an routing and spectrum allocation (RSA) algorithm is to achieve the best performance within the limits set by the physical constraints. Recently, an increasing number of studies have investigated solutions to the RSA problem in the EON. The RSA problem can be cast in numerous forms. This chapter discusses various issues related to RSA and sliceable bandwidth-variable transponder (SBVT) as follows.

5.1 Fragmentation

EONs allocate spectrum on contiguous subcarrier slots. As the size of contiguous subcarrier slots is elastic, it can be a few GHz or even narrower. Therefore, dynamically setting up and tearing down connections can generate the bandwidth fragmentation [31, 111] problem. It is the condition where available slots become isolated from each other by being misaligned along the routing path or discontiguous in the spectrum domain. Thus, it is difficult to utilize them for upcoming connection requests. If no available slot can fulfill the required bandwidth demand of a connection request, the connection request is considered to be rejected/blocked. This is called call blocking and drives network operators to

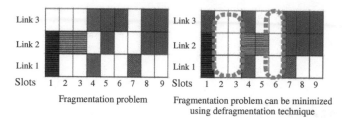

Figure 5.1: Fragmentation problem, and reducing its effect by incorporating defragmentation technique.

periodically reconfigure the optical paths and spectrum slots. This is referred to as network defragmentation. Figure 5.1 shows the fragmentation problem.

To overcome the bandwidth fragmentation problem, many RSA approaches [29, 31, 69, 78, 112] have been published. In this direction, Kadohata et al. [67] and Zhang et al. [68] developed bandwidth defragmentation schemes by considering the green field scenario, where connections are totally rerouted. Patel et al. [113] formulated the defragmentation problem in an EON as an ILP formulation to provide optimal defragmentation with consideration of the spectrum continuity and contiguity constraints. They presented two heuristic algorithms, namely—(i) greedy-defragmentation algorithm and (ii) shortest-path-defragmentation for large-scale networks in order to maximize the spectrum utilization. Fragmentation-aware RSA algorithms or defragmentation approaches can be classified into two categories, namely—(i) proactive fragmentation-aware RSA and (ii) reactive fragmentation-aware RSA, which are discussed in Section 5.1.1 and Section 5.1.2, respectively.

5.1.1 Proactive fragmentation-aware RSA

When a new request is admitted to the network, the proactive fragmentation-aware RSA technique attempts to prevent or minimize spectrum fragmentation in the network. In this direction, Wang et al. [99] have presented four spectrum management techniques for allocating spectrum resources to connections of different data rates. In their approaches, all connections share the whole spectrum using the first-fit spectrum allocation policy. A similar concept of spectrum reservation has been presented by Christodoulopoulos et al. [31]. In their approach, a block of contiguous subcarriers is reserved for each source-destination pair. In addition, subcarriers that are not reserved may be shared among all connections on demand.

5.1.2 Reactive fragmentation-aware RSA

In a dynamic environment, the fragmentation problem cannot be completely eliminated. Therefore, reactive fragmentation-aware RSA algorithms attempt to restore the network's ability to accommodate high-rate and long-path connections. The main objective of defragmentation is to reconfigure the spectrum allocation of existing connections in order to consolidate available slots into large contiguous and continuous blocks that may be used for upcoming connection requests. In this direction, Wang et al. [99] have presented a set of reactive defragmentation strategies that exploit hitless optical path shift (HOPS). HOPS technology shifts a connection to a new block of spectrum as long as the route of the connection does not change and the movement of the new spectrum does not affect other established connections. They presented a scheme that consolidates the spectrum slots freed by a terminated connection with other blocks of spectrum available along the links of its path.

Most of the approaches [29, 31, 69, 78, 99, 111, 112] in the literature perform bandwidth defragmentation after bandwidth fragmentation occurs in subcarrier-slots. This means that the traffic is disrupted by connection rerouting, as the bandwidth defragmentation is performed. Bandwidth defragmentation that uses connection rerouting increases the traffic delay and system complexity.

To overcome this serious issue, R. Wang and B. Mukherjee [114] have presented a scheme that prevents bandwidth fragmentation without performing any connection rerouting. Typically, when connection requests with lower-bandwidth and higher-bandwidth are not separated during spectrum allocation, it is more likely that the higher-bandwidth connection requests are blocked. In order to circumvent this drawback, they explore an admission control mechanism that captures the unique challenges posed by heterogeneous bandwidths. They adopt a preventive admission control scheme based on spectrum partitioning to achieve higher provisioning efficiency. As a result, it prevents the blocking of connections due to the unfairness of bandwidth issue.

Similarly, Fadini et al. [100] have presented a subcarrier-slot partition scheme for spectrum allocation in EONs; it reduces the number of non-aligned available slots without connection rerouting. Thus the blocking probability in the network is reduced. In their approach, they define a connection group as a set of connections whose routes are exactly the same. When the spectrum resources of two connections from different connection groups sharing a common link are allocated to adjacent slots, some available slots might be non-aligned with each other. When another connection request arrives in the network and its route needs these available slots, the connection request is rejected if these available slots are non-aligned. The partitioning scheme divides the subcarrier slots into several partitions. Disjoint connections whose routes do not share any link are allocated to the same partition, while non-disjoint connections are allocated to different partitions. In this way, their approach increases the number of aligned available slots in the network and hence the blocking probability is reduced.

5.2 Modulation

The traditional WDM-based optical network assigns spectrum resources to optical paths without considering the appropriate modulation technique, which leads to an inefficient utilization of the spectrum. However, the OFDM-based EON allocates optical paths with consideration of adaptive modulation and bit rate to further improve the spectrum efficiency. In the modulation-based spectrum allocation scheme [29, 115], the necessary minimum spectral resource is adaptively allocated to an optical path. The adaptation considers the physical conditions while ensuring a constant data rate. The modulation-based spectrum allocation scheme improves the spectrum efficiency, as the allocated spectral bandwidth can be reduced for shorter paths by increasing the number of modulated bits per symbol.

In this direction, Jinno et al. [29] have presented a distance-adaptive spectrum allocation scheme that adopts a high-level modulation format for long distance paths, and a low-level modulation format to shorter paths. As the optical signal-to-noise ratio (OSNR) tolerance of 64-QAM is lower than that of QPSK, it suits shorter distance lightpaths as shown in Fig. 5.2.

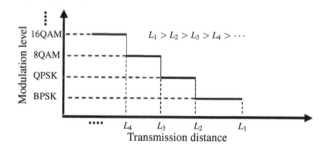

Figure 5.2: Modulation level versus transmission distance.

The modulation based spectrum allocation schemes can be classified into two categories, namely—(i) offline modulation based spectrum allocation, and (ii) online modulation based spectrum allocation, and are discussed below.

5.2.1 *Offline modulation based spectrum allocation*

Christodoulopoulos et al. [31] have presented an offline modulation based spectrum allocation scheme, where a mapping function is provided as input to the problem. In their scheme, each demand is mapped to a modulation level according to the requested data rate and the distance of the end-to-end path. Initially, they presented a path-based ILP formulation for their scheme, and then decomposed the problem into two sub-problems, namely (i) routing and modulation level (RML), and (ii) spectrum allocation. They solved the sub-problems sequentially using ILPs. Finally, a sequential algorithm was presented to serve connections one-by-one, and to solve the planning problem sequentially.

5.2.2 *Online modulation based spectrum allocation*

Most of the studies on online modulation based spectrum allocation [116–119] have introduced heuristic algorithms, which deal with randomly arriving connection requests. Initially, these algorithms compute a number of fixed-alternate paths for each source-destination pair, and arrange them in decreasing order of their end-to-end path length. In the second step, a spectrum allocation policy is used to allocate a lightpath to each connection request by considering alternate path routing and modulation.

Recent studies [32, 116, 117] on modulation based spectrum allocation claim that this type of spectrum allocation scheme increases the spectral utilization by approximately 9%-60% compared to fixed-modulation based spectrum allocation in the EON. Fixed-modulation based spectrum allocation schemes do not consider the most appropriate modulation technique for different connection requests according to their lightpath distance. Typically, they select, conservatively, one modulation technique for all connection requests regardless of their lightpath distance. As an example, a fixed-modulation based spectrum allocation scheme adopts the BPSK modulation format for all connection requests regardless of their lightpath distance. As a result, this type of spectrum allocation schemes does not utilize the spectrum efficiently. On the other hand, modulation-based spectrum allocation schemes determine the modulation technique that best suits each lightpath distance. As an example, a modulation based spectrum allocation scheme adopts BPSK for long distance lightpaths, and 16-QAM for shorter distance lightpaths. This minimizes the number of spectrum slots that must be assigned, which yields better utilization of spectrum resources compared to fixed-modulation based spectrum allocation schemes.

5.3 Quality of transmission

The EON architecture offers the ability to choose the modulation format and channel bandwidth to suit the transmission distance and quality of transmission desired. One version of the online modulation based spectrum allocation scheme is referred to as quality-of-transmission aware RSA. In this direction, Beyranvand et al. [119] have presented a quality of transmission (QoT) aware online RSA scheme for the EON. Their approach employs three steps, namely—(i) path calculation, (ii) paths election, and (iii) spectrum assignment to construct the complete framework. The Dijkstra and k-shortest path algorithms have been adapted for computing paths, while fiber impairments and non-linear effects on the physical layer are modeled to estimate the QoT along the given route. Yang et al. have presented [118] a QoT-aware RSA scheme in order to select a feasible path for each requested connection and allocate subcarrier slots by using modulation formats appropriate for the transmission reach and requested data rate.

5.4 Traffic grooming

In WDM-based wavelength-routed optical networks, traffic grooming [6, 22, 120–126] is used to multiplex a number of low-speed connection requests into a high-capacity wavelength channel for enhancing channel utilization. Traffic grooming improves the resource utilization by aggregating multiple electrical channels (packet or circuit flows) onto one optical channel.

The EON allocates spectral resources with just enough bandwidth to satisfy the traffic demands. However, traffic grooming [58, 123, 127–129] is still essential for the following reasons, (i) BVT is normally designed so as to maximize the traffic rate in the network, and it does not support slicing at a very early stage [130]. Electrical traffic grooming is applied in order to use transponder capacity efficiently. (ii) Generally speaking, a filter guard band between two adjacent channels should be assigned to resolve optical filter issues. Traffic grooming can minimize filter guard band usage by aggregating traffic electrically. The electrical switching fabric is still needed for traffic grooming in the EON, similar to WDM networks. The main difference is that the transponder used in EONs does not strictly follow the ITU-T central frequency. As a result, it can provide flex-optical channels.

To further improve the flexibility and eliminate the electrical processing, researchers have designed SBVT for the EON. In SBVT-based EONs, the traffic grooming [58, 131] function can be partly offloaded from the electrical layer to the optical layer. Multiple electrical channels are groomed onto one sub-transponder channel, in which each sub-transponder channel is associated with a flex-optical channel. Multiple sub-transponder channels (flex-optical channels) are groomed optically onto one transponder by using an optical switching fabric (e.g., BV-OXC), which is called optical traffic grooming. Figure 5.3 distinguishes between traffic grooming in WDM-based optical networks, traffic grooming with BVT in EONs, and traffic grooming with SBVT in EONs. Spectrum efficiency and transponder usage are improved in WDM-based optical networks, whereas traffic grooming with BVT in the EON improves transponder usage and reduces guard band usage. Finally, traffic grooming with SBVT in the EON eliminates electrical processing by offloading parts of the grooming function to the optical layer.

Zhang et al. [127] have incorporated, for the first time, a grooming approach for the RSA problem in the EON with BVT. In their approach, multiple low-speed connection requests are groomed into elastic optical paths by using electrical layer multiplexing. They presented a mixed integer linear pro-

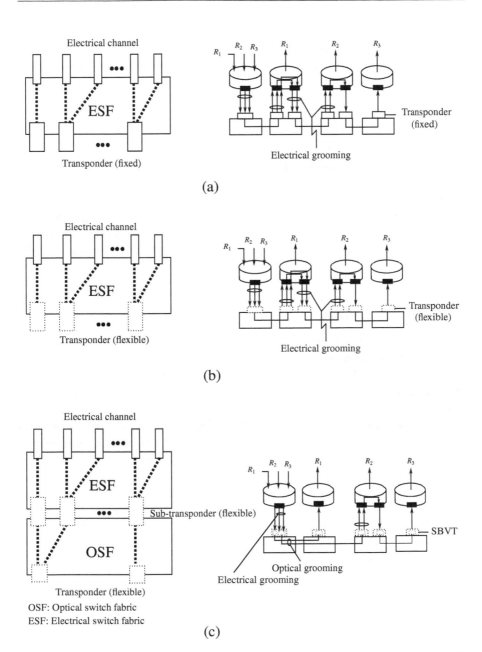

Figure 5.3: Comparison of traffic grooming in (a) WDM-based optical networks, (b) elastic optical networks with BVTs, and (c) elastic optical networks with SBVTs.

gramming (MILP) formulation to reduce the average spectrum utilization in the traffic-grooming scenario. Zhang et al. [123] have presented a multi-layer auxiliary graph to implement various traffic-grooming policies by properly adjusting the edge weights in the auxiliary graph. With their approach, they have shown that there is a trade-off among different traffic-grooming policies, and that the spectrum reservation scheme can be incorporated into various traffic-grooming scenarios. Recently, Zhang et al. [58] have presented dynamic traffic grooming in SBVT-enabled EONs. In their approach, a three-layered auxiliary graph (AG) model has been presented to address mixed-electrical-optical grooming under the dynamic traffic scenario. By adjusting the edge weights of AG, various traffic-grooming policies can be achieved for different purposes. Furthermore, two spectrum reservation schemes have been introduced in order to efficiently utilize transponder capacity. Finally, they compared different traffic-grooming policies under two spectrum reservation schemes, and the tradeoff among the policies was shown.

5.5 Survivability

The EON has the capability to support individual data rates from 400-1000 Gb/s [132]. It also aggregates the throughput per fiber link to approximately 10-100 Tb/s. Therefore, failure of a network component, such as optical fiber or network node, can disrupt communications for millions of users, which can lead to a great loss of data and revenue. As an example, in 2004, the Gartner Research Group had lost approximately 500 million dollars due to failure of a optical network [133]. Thus, survivability against failure has become an essential requirement of the EON. Failure recovery [134] is defined here as "*the process of re-establishing traffic continuity in the event of a failure condition affecting that traffic, by re-routing the signals on diverse facilities after the failure*". A network is defined as survivable [134], if its recovery can be secured rapidly. Similar to WDM-based optical networks, the survivability mechanisms [89, 90, 135] for the EON can be classified into two broad categories, namely—(i) protection and (ii) restoration, which are discussed briefly in the following subsections.

5.5.1 Protection

The protection techniques of [135–139] use backup paths to carry optical signals after fault occurrence. The backup paths are computed prior to fault occurrence, but they are reconfigured after fault occurrence. In this direction, Klinkowski et al. [139] have presented an RSA approach with dedicated protection for static traffic demands. Although dedicated protection can provide more reliability, it is unable to utilize the spectrum slots properly as some of the spectrum slots are reassigned prior to fault occurrence. To overcome this problem, Liu et al.

[135] presented a shared protection scheme to enhance the spectrum utilization by sharing backup spectrum slots between two adjacent paths on a link, if the corresponding working paths are link-disjoint. They explored the opportunity of sharing enabled by tunable transponders. The elasticity of the transponder enables the expansion and contraction of paths. As a result, the backup spectrum is used by only one of the adjacent paths at a time. Similarly, Shen et al. [136] addressed a shared protection technique for EONs to minimize both required spare capacity and maximum number of used link spectrum slots.

5.5.2 Restoration

In restoration [90, 140–145], backup paths are computed dynamically based on the latest network information after fault occurrence, and hence can provide more efficiency in terms of resource utilization compared to protection. In this direction, Ji et al. [146] presented three algorithms for dynamic preconfigured-cycle (p-cycle) configuration in order to provide the EON with 100% restoration against single-link failure. The first algorithm configures the working path and p-cycles of a request together according to the protection efficiencies of the p-cycles. In order to reduce the blocking probability, they presented a spectrum planning technique that regulates the working spectrum and protection resources, and finally, two algorithms based on the protected working capacity envelope cycles and Hamiltonian cycles.

Paolucci et al. [145] have presented a restoration technique enabling multi-path recovery and bit rate squeezing in the EON. They exploited the advanced flexibility provided by sliceable bandwidth-variable transponders that support the adaptation of connection parameters in terms of the number of sub-carriers, bit rate, transmission parameters, and reserved spectrum resources. They formulated their problems as an ILP model and finally, presented an heuristic algorithm that efficiently recovers network failures by exploiting limited portions of the spectrum resources along multiple routes. As restoration finds the backup paths after fault occurrence, it offers slower recovery than protection. Depending on the type of rerouting used, restoration can be considered as consisting of three categories, namely—(i) link restoration, (ii) path restoration and (iii) segment-based restoration. Link restoration discovers a backup path for the failed connection only around the failed link. In path restoration, the failed connection independently discovers a backup path on an end-to-end basis. Segment-based restoration discovers a segment backup path of the failed connection.

5.6   Energy saving

The energy consumption of telecom networks is drastically increasing with the increase in traffic. The IP router consumes the maximum amount of energy in IP-over WDM-based optical networks [147]. When transmission rate increases, the optical transponder associated with the IP router is a huge energy consumer in optical networks [148]. Therefore to minimize energy consumption, it is essential to reduce the number of IP router ports and optical transponders. By using the advantages of SBVT, the EON can offer some new features for traffic grooming (shown in Fig. 5.3) and optical layer bypass, which can help to reduce the energy consumption. In this direction, Zhang et al. [149] studied the power consumption of IP-over-EONs with different elastic optical transponders. The results of their studies show that significant energy savings are possible if SBVT is used rather than the fixed BVT.

Recently, researchers [150–152] have focused on energy efficient RSA schemes for the EON. In this direction, A. Fallahpour et al. [151] have presented a dynamic energy efficient RSA algorithm that considers regenerator placement to suppress the total network energy consumption. In their work, a virtual graph is designed based on the given network topology, where the cost functions of the virtual graph are computed according to the energy consumption of the corresponding links and intermediate routers. Furthermore, a newly arrived connection request is served by finding the most energy-efficient path among the possible candidate paths. Similarly, Zhang et al. [152] have presented energy-efficient dynamic provisioning in order to significantly reduce energy consumption and make efficient use of spectrum resources. In their research work, they adopt an auxiliary graph, and from it created a dynamic provisioning policy called time-aware provisioning with bandwidth reservation (TAP-BR). The TAP-BR policy incorporates the two important factors of time awareness and bandwidth reservation, in order to facilitate energy-efficient provisioning.

Several studies on energy saving in the EON are anticipated, and more research work is needed to develop truly effective energy-saving RSA schemes.

5.7   Networking cost of SBVT

This subsection discusses the networking cost reduction made possible by the use of SBVTs in the EON. We have observed from the literature that SBVTs allow the reuse of hardware and optical spectrum by transmitting data to multiple destinations. SBVTs enable point to multiple point transmission where the traffic rate to each destination and the number of destinations can be freely set to satisfy the request. On the other hand, the non-sliceable transponder requires at least one interface for each destination, which increases networking cost.

In this direction, López et al. [57, 59] have presented two node models, namely—(i) non-sliceable transponder model, and (ii) SBVT model in order to

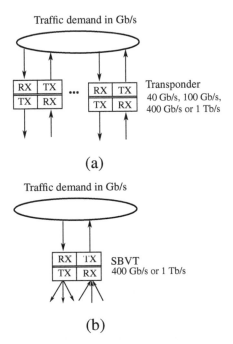

Figure 5.4: Different models (a) non-sliceable transponder model, and (b) SBVT model for the analysis of networking cost.

compare the networking cost, please see Fig. 5.4. The main difference between these two models is that the non-sliceable transponder model requires at least one interface for each destination, while the SBVT transponder reuses hardware and optical spectrum to transmit to multiple destinations. The model without sliceable transponders considers coherent modulation formats, such as—40 Gb/s, 100 Gb/s, 400 Gb/s and 1 Tb/s, whereas only 400 Gb/s or 1 Tb/s SBVTs are considered by the SBVT model.

Finally, the result of the research work [57] has claimed that using 400 Gb/s and 1 Tb/s SBVTs reduces transponder costs in the network by at least 50% in a core network scenario. This reduction was calculated relative to BVTs of 400 Gb/s and 1 Tb/s in the non-sliceable scenario.

A significant number of works have addressed various aspects of the RSA problem in the EON. Table 5.1 summarizes these different aspects, namely—fragmentation, modulation, quality-of-transmission, traffic grooming, survivability, energy saving, and networking cost reductions by SBVT.

Table 5.1: Summaries of different issues related to RSA.

Various issues		Reference
Fragmentation	Proactive	[31],[100, 114]
	Reactive	[29, 31, 69, 78, 99, 111, 112]
Modulation	Online	[116–119]
	Offline	[31]
Quality of transmission		[118, 119]
Traffic grooming		[123, 127, 128, 145],[58, 129]
Survivability	Protection	[135–138]
	Restoration	[90, 140–144],[146]
Energy saving		[151],[150, 152]
Networking cost caused by SBVT		[57, 59]

Exercises

1. Must proactive and reactive fragmentation-aware RSA approaches be disjoint? Justify your answer.

2. What are the differences between protection and restoration techniques? Discuss the pros and cons of protection and restoration techniques.

3. How does EON offer better energy efficiency over WDM based optical networks?

4. How does SBVT reduce cost in the network over BVT?

5. What is the relationship between transmission reach and modulation technique?

6. Fragmentation has a negative impact in EONs. Justify this statement.

Chapter 6

Spectrum Fragmentation Problems in Elastic Optical Networks

Spectrum fragmentation, is a serious issue in elastic optical networks (EONs), which suppresses the resource utilization and enhances call blocking in the network. In EONs, lightpaths are established and released dynamically, which causes bandwidth fragmentation that occurs when unoccupied isolated slots are not aligned along the route and contiguous in the spectrum. This chapter presents the fragmentation problem in EONs. We discuss the different metrics used to measure the spectrum fragmentation for an EON link, and compare them in terms of their pros and cons. We further present how to estimate the fragmentation of an entire network. The major spectrum allocation approaches, which are random fit, last fit, first fit, first-last fit, least used, most used, and exact fit, are analyzed in terms of overall network fragmentation.

6.1   Overview of spectrum fragmentation

The spectrum fragmentation problem [31, 68, 94] in EONs, where lightpath requests are allocated dynamically while respecting the constraints of spectrum continuity and contiguity. The spectrum continuity constraint [6] ensures that all hops in the end-to-end route of a lightpath use the same spectrum slots. In each hop, a lightpath can use more than one spectrum slot, but these spectrum slots must neighbor each other, which is known as the constraint of spectrum contiguity [72]. We have already explained the concept of spectrum continuity and contiguity constraints in Chapter 4.

When EONs consider dynamic traffic, lightpath requests arrive for lightpath provisioning and releasing in the network at any time. Dynamically provisioning and releasing lightpaths will trigger the bandwidth fragmentation problem [31, 68, 69, 94, 153]. This problem happens when vacant isolated slots are either not aligned along the route or not contiguous in the spectrum. When one or more vacant slots of dissimilar links on the lightpath route are not the same, non-aligned vacant slots occur. Non-contiguous vacant slots are created in the spectrum domain when one or more vacant slots are not adjacent to each other. It is highly unlikely that these non-aligned and non-contiguous vacant slots can be used to satisfy future lightpath requests. When the required slots of a lightpath request are not satisfied, the lightpath request is disallowed, which causes call blocking in the network.

Figure 6.1 explains how fragmentation can yield call blocking in a network. We consider a lightpath request that demands two slots, as shown in Fig. 6.1(a). Although two free slots are available in the network, the lightpath request can not be established. This is because these slots are either not aligned along the route (see Fig. 6.1(b)) or they are not adjacent in the spectrum domain (see Fig. 6.1(c)).

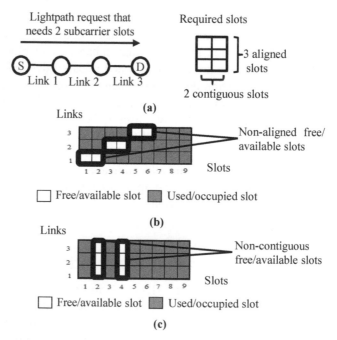

Figure 6.1: (a) Slot requirement for spectrum allocation, (b) fragmentation due to non-aligned free/available slots, and (c) fragmentation due to non-contiguous free/available slots.

6.2 Fragmentation metrics

This section describes the different metrics that are used to measure spectrum fragmentation in EONs. Spectrum loss can be a metric to measure fragmentation; it occurs when a free slot cannot be utilized. Spectrum loss is demand sensitive as larger granularities are more likely to be blocked than smaller granularities. Therefore, it is not considered to estimate the level of fragmentation effect in EONs; other metrics are essential. The most commonly used metric to estimate fragmentation is the blocking ratio; it is the ratio of the number of blocked requests to the number of offered requests in the network. The assumption is that if the blocking ratio is less, the fragmentation effect is also less. However, the blocking ratio is not a complete measure of fragmentation as the blocking ratio is also impacted by several system parameters, such as a lack of resources, quality of transmission, and holding time. Therefore, it is necessary to identify other comparison metrics to measure the blocking caused by spectrum fragmentation.

6.2.1 Measuring fragmentation in a link

In the following, we discuss the metrics that are used to estimate spectrum fragmentation in each link. Let i be a block of available contiguous slots, f_i be the number of available contiguous slots in block i, and I be a set of blocks of available contiguous slots, in each link. In the case that there is one slot whose neighbor slot(s) are not available, we also consider it as a block $i \in I$ with $f_i = 1$. We express $A = \max_{i \in I} f_i$ and $B = \sum_{i \in I} f_i$.

6.2.1.1 External fragmentation metric

External fragmentation metric [154, 155] has been well studied in memory fragmentation management. It can also be used to measure the fragmentation effect in each link, which is defined by (6.1)

$$\Theta = 1 - \frac{A}{B}, \tag{6.1}$$

where A and B are the maximum number of available contiguous slots, and the number of all available slots, respectively, in each link. The concept of the external fragmentation metric is explained by the example shown in Fig. 6.2.

6.2.1.2 Entropy-based fragmentation metric

In information theory, the amount of information contained in a message can be measured by entropy [156]. Taking this direction, Wright et al. [157] used the concept of entropy as a quantitative metric for measuring spectrum fragmentation. The entropy-based fragmentation metric is defined by (6.2)

$$\tau = \sum_{i \in I} \frac{f_i}{S} \ln\left(\frac{S}{f_i}\right), \tag{6.2}$$

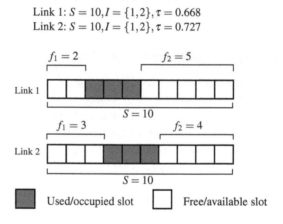

Link 1: $I = \{1,2\}, A = \max(2,5) = 5, B = 2+5 = 7, \Theta = 0.285$
Link 2: $I = \{1,2\}, A = \max(3,4) = 4, B = 3+4 = 7, \Theta = 0.428$

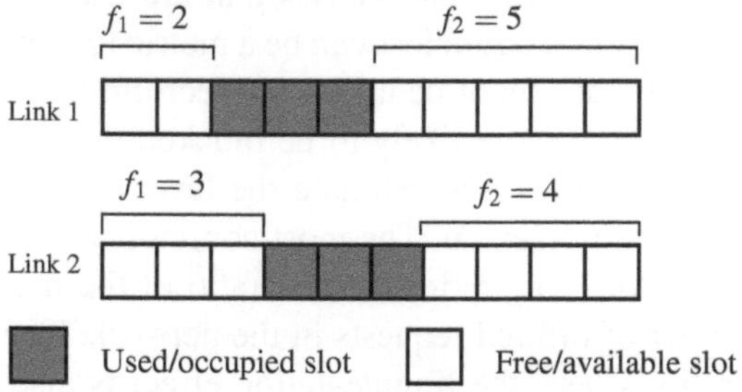

Figure 6.2: Illustration of external fragmentation metric.

Link 1: $S = 10, I = \{1,2\}, \tau = 0.668$
Link 2: $S = 10, I = \{1,2\}, \tau = 0.727$

Figure 6.3: Illustration of entropy-based fragmentation metric.

where S denotes the total number of slots. The concept of the entropy-based fragmentation metric is explained by an example, as shown in Fig. 6.3.

6.2.1.3 Access blocking probability metric

The access blocking probability metric [158] can be used to estimate fragmentation. In determining the access blocking probability metric, it is considered that blocking is dependent only on the granularities used by the transponders. The access blocking probability metric is defined by (6.3)

$$\sigma = 1 - \frac{\sum_{i \in I} \sum_{k=1}^{w} \mathrm{DIV}(f_i, G_k)}{\sum_{j=1}^{w} \mathrm{DIV}(B, G_j)}, \tag{6.3}$$

where w denotes types of granularities used by transponders. B represents the number of all available slots. G_k represents the number of slots required to sat-

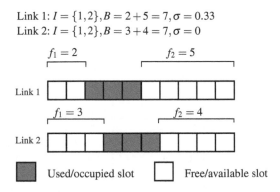

Link 1: $I = \{1,2\}, B = 2+5 = 7, \sigma = 0.33$
Link 2: $I = \{1,2\}, B = 3+4 = 7, \sigma = 0$

Figure 6.4: Illustration of access blocking probability metric.

isfy the granularity type k. $\mathrm{DIV}(a,b)$ indicates the integer division of a/b. The concept of the access blocking probability metric is explained by the example shown in Fig. 6.4. In this example, three and four slots are considered to satisfy the two types of granularities (i.e., $G_1 = 3$ and $G_2 = 4$).

6.2.1.4 Comparing different link-oriented fragmentation metrics

This subsection compares the different metrics considered for estimating the degree of fragmentation. The different link-oriented fragmentation metrics are summarized in Table 6.1. The external fragmentation metric has the lowest time complexity of all considered metrics as it inspects only the largest block. It recognizes that there is no fragmentation effect if all free slots are contiguous. In this case, the maximum number of available contiguous slots is equivalent to the number of slots in the link, which indicates the initial condition of the link. The disadvantage of the external fragmentation metric is that it ignores the small fragments to focus on the maximum number of available contiguous slots in the link.

Similar to the external fragmentation metric, the entropy-based fragmentation metric is unable to distinguish the case when fragmented slots match the available granularities from the inappropriate fragmented cases. This metric can efficiently estimate relative fragmentation, and so allows comparisons of different arrangements. The time complexity of the entropy-based fragmentation metric is linear to the number of fragments.

The access blocking probability metric can estimate fragmentation more comprehensively than the previous two metrics. When all fragments are smaller than the smallest granularity, the access blocking probability metric finds that the spectrum is completely fragmented. If the spectrum is not fragmented, it returns zero, which indicates that all slots are free and contiguous. As the access blocking probability metric returns a relative value between zero and one, it can

Table 6.1: Summaries of different link-oriented fragmentation metrics.

Fragmentation metrics	Reference	Time complexity	Observations and comments
External fragmentation	[154, 155]	Lowest	It ignores the small fragments as it focuses on the maximum number of available contiguous slots in the link.
Entropy-based fragmentation	[156]	Moderate	It considers any fragmented slot and estimates relative fragmentation.
Access blocking probability	[158]	Highest	It deals with available granularities and tries to avoid situations where these granularities are blocked.

also be used to compare relative levels of fragmentation; the more the spectrum is fragmented, the greater the relative fragmentation is. However, similar to the external fragmentation metric, the access blocking probability metric cannot differentiate whether (i) all slots are free, (ii) all slots are used, and (iii) all free slots are contiguous. This metric has higher time complexity than the other metrics considered here.

To accurately estimate the effect of fragmentation in a single fiber link, some works that use Markov Chain (MC) models [159, 160] can be found in the literature. Taking this direction, Kim et al. [159] presented an MC model that attempts to characterize the fragmentation problem in a single fiber link. In addition, their model is able to accurately capture the blocking probability due to the fragmentation effect. The authors in [160] introduced an exact MC model for single-link flexible grid networks and analyzed the blocking probability caused by fragmentation.

We summarize that the above metrics are considered to estimate the fragmentation in each link. However, it is difficult to estimate the overall fragmentation in a network due to the spectrum continuity and contiguity constraints. In the next subsection, we define how fragmentation in a network can be estimated.

6.2.2 Measuring fragmentation in a network

The fragmentation in a network can be measured with the help of contiguous-aligned available slot ratio [161]. We use the routes of source-destination pairs to represent the contiguous aligned available slot ratio in the network. The contiguous-aligned available slot ratio is defined by $\phi = \sum_{d \in D} \sum_{k \in K_d} w_{dk} \cdot \psi_{dk}$, where $\psi_{dk} = \frac{\gamma_{dk}}{Z}$. w_{dk} is a weight that is proportional to the traffic load of route $k \in K_d$ of source-destination pair $d \in D$, where $\sum_{d \in D} \sum_{k \in K_d} w_{dk} = 1$. Z represents the total number of spectrum slots in each link; we assume that all links have the same number of slots. D and K_d represent the set of all the source-destination pairs and the set of routes of source-destination $d \in D$, respectively.

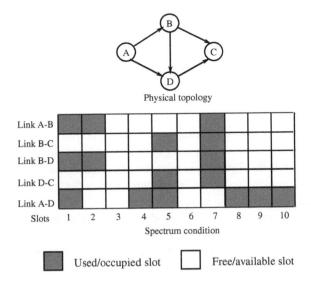

Figure 6.5: Illustration of fragmentation in a network.

γ_{dk} represents the maximum number of contiguous aligned available slots for route $k \in K_d$ of source-destination pair $d \in D$. The fragmentation is defined by $\chi = 1 - \phi$.

The fragmentation in a network is illustrated in Fig. 6.5. We assume four-node network with five directed links. Each link consists of 10 spectrum slots. We assume six source-destination pairs, which are AB $(d = 1)$, AC $(d = 2)$, AD $(d = 3)$, BC $(d = 4)$, BD $(d = 5)$, and DC $(d = 6)$. The routes of AB, AC, AD, BC, BD, and DC are A-B, A-B-C, A-B-D, B-C, B-D, and D-C, respectively. Here, $Z = 10$ and $|D| = 6$. For the sake of simplicity, we assume that each source-destination pair has one route. The traffic load and the average number of requested slots for each source-destination pair is the same over all the source-destination pairs. Therefore, with $|K_d| = 1$ for all $d \in D$, $w_{d1} = 1/6$ is set. We obtain $\gamma_{11} = 4, \gamma_{21} = 6, \gamma_{31} = 8, \gamma_{41} = 4, \gamma_{51} = 4$, and $\gamma_{61} = 4$. The contiguous-aligned available slot ratio and the fragmentation in the network are 0.5 and 0.5, respectively.

6.3 Impact of fragmentation on spectrum allocation policies

This section discusses the major spectrum allocation policies, which are first fit, random fit, last fit, first-last fit, exact fit, least used, most used, and presents the impact of fragmentation when the network uses these spectrum allocation policies. We assume that lightpath requests arrive in the system sequentially, and they are established one by one if they share the same link. More than one request can be established in parallel if they do not share any common link. To evaluate the fragmentation effect, we use the contiguous-aligned available slot ratio metric due to its simplicity.

6.3.1 First fit

The first fit spectrum allocation policy [98, 99] indexes all the spectrum slots and maintains a list of indexes of available and used spectrum slots. For each allocation, it tries to select the lowermost indexed available slot and use it for lightpath provisioning. When the lightpath is released, the used slots are added to the list of freed slots. By choosing spectrum in this way, lightpaths are crammed into fewer spectrum slots, which helps to increase the contiguous-aligned available slot ratio in the network. Employing this policy does not require any global information of the network. This policy is recognized as suitable for spectrum allocation. This is because it provides higher contiguous-aligned available slot ratio and its computation time complexity is low.

6.3.2 Random fit

The random fit policy [6, 98] maintains a list of free spectrum slots. When a lightpath request arrives in the network, the policy arbitrarily chooses an available spectrum slot and uses it for lightpath provisioning. After slots are assigned, the list of free slots is refreshed by removing the just-assigned slot from the available list. Once a lightpath is released, the just-released slot is added to the list of free slots. By selecting spectrum slots in a random manner, the network operator tries to reduce the possibility that some specific spectrum slots are often used. In this case, allocated spectrum slots are uniformly distributed over entire spectra.

6.3.3 Last fit

The last fit allocation policy [6, 100] invariably tries to select the highest indexed available slot and use it for lightpath provisioning. When the lightpath is released, the slot is added to the list of free slots. The last fit spectrum allocation policy matches the computational complexity of the first fit spectrum allocation policy. This policy is not suitable for transmission systems that do not adopt dispersion compensation including digital signal processing. If we do not adopt dispersion compensation, the overall dispersion effect of a lightpath using last fit is higher than that of using first fit. This is because the dispersion effect increases with increase in the wavelength range. As an example, the dispersions of the wavelengths of 1.52 μm and 1.53 μm are 18 ps/nm/km and 19.5 ps/nm/km, respectively.

To overcome the problem of last fit policy, the work in [109] suggests that longer lightpath requests need to be assigned from the smallest indexed spectrum slot, where the dispersion effect is less, and shorter lightpath requests need to be assigned from the largest indexed spectrum slot, where the dispersion effect is more. As a result, a less robust modulation technique is used to maintain the QoT threshold level for the longer lightpath requests, and hence a lower number of spectrum slots are required for lightpath establishment.

6.3.4   First-last fit

The first-last fit spectrum allocation policy [161, 162] is a combination of first fit and last fit spectrum allocation policies. Lightpath requests in the network are divided into two groups; one group uses the first fit allocation policy and the other group uses the last fit spectrum allocation policy. The group formulation can be performed in several ways. In [161], lightpath requests are grouped based on disjoint and non-disjoint paths. Lightpaths with disjoint paths are allocated using the first fit allocation policy, whereas lightpaths with non-disjoint paths are allocated using the last fit allocation policy. The intention of this policy is to avoid spectrum fragmentation in the network. The performance of this policy depends mainly on how lightpath groups are formed; if the lightpath groups are not properly formed, its performance is degraded [100].

6.3.5   Least used

The free spectrum slots, which have been utilized by the smallest number of fiber links in the network, are the focus of the least used spectrum allocation policy [6, 7] to satisfy lightpath requests. If several spectrum slot candidates are equally possible, the first fit spectrum allocation policy is used to select the best candidate. Choosing spectrum slots in this way attempts to distribute the load uniformly across the entire spectrum domain.

6.3.6   Most used

The spectrum allocation of the most used policy [6,7] is similar to that of the least used policy. Instead of choosing the spectrum resources that have been utilized by the least number of fiber links in the network, this policy selects free slots, which have been utilized by the most number of fiber links in the network. Choosing spectrum slots using this policy is an effort to enhance the reuse of spectrum in the network.

6.3.7   Exact fit

If a lightpath request arrives, which requires c contiguous slots for lightpath establishment, the exact fit allocation policy [98, 161] finds c contiguous free slots from the beginning of the spectrum. If there exists exact c free slots, which are not more than or less than c, it allocates those slots. Otherwise, this spectrum allocation follows the first fit spectrum allocation policy. The exact fit allocation

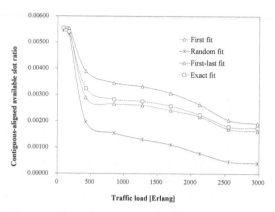

Figure 6.6: Comparison of major spectrum allocation policies in terms of contiguous-aligned available slot ratio in NSFNET [97].

policy tries to suppress the fragmentation effect and improves the contiguous-aligned available slot ratio in the network.

Several studies in the literature have already been conducted to analyze the effect of fragmentation using different spectrum allocation policies [97, 161, 162]. Figure 6.6 condenses the major spectrum allocation policies. We compare the performance of these spectrum allocation policies in terms of contiguous-aligned available slot ratio under varied traffic load (in Erlangs); the traffic load in the network is defined in Chapter 7 (page number 85). We estimate the relationship between the traffic load and the average slot utilization in NSFNET [6]. The assumptions, including lightpath request distribution and maximum number of occupied spectrum by lightpaths, of this evaluation are explained in Chapter 7 (page number 85). We observe from the literature that the least-used and most-used allocation policies have the highest time complexity [6]. Note that the last fit allocation policy is the same as with the first fit policy, in terms of the contiguous-aligned available slot ratio. We obtain the numerical results via simulation performed on NSFNET. We observe that, when the traffic load is low, the contiguous-aligned available slot ratios are comparable for different schemes. The numerical results suggest that the first-last fit and random fit policies yield the highest and lowest contiguous-aligned available slot ratios, respectively, among all spectrum allocation policies, when the traffic load is reasonably high. In other words, we can say that the first-last fit allocation policy offers the lowest fragmentation effect among the spectrum allocation policies considered here. When the traffic load increases more and the network becomes congested, the decrease rate of contiguous-aligned available slot ratio becomes slow with traffic load for any allocation policy.

The above spectrum allocation policies can use any of the routing policies [6, 80, 81], namely (i) fixed routing, (ii) fixed alternate routing, (iii) least congested routing, and (iv) adaptive routing, to find routes between source-destination pairs. However, the performance of a spectrum allocation policy depends on the selection of a routing policy. The selection of routes can be performed for lightpath requests either before spectrum allocation or during the spectrum allocation.

Fixed routing (FR) [6, 80] precomputes a single fixed route for each source-destination pair using some shortest path algorithms, such as Dijkstra's algorithm [85]. When a lightpath request arrives in the network, it tries to establish a lightpath along the precomputed fixed route. FR checks whether the desired slot is free on each link of the precomputed route or not. The lightpath request is blocked if one link of the precomputed route does not have the desired slot.

In fixed alternate routing (FAR) [6, 80], an ordered list of a number of fixed routes for each source-destination pair is maintained in the network. These routes are determined off-line. When a lightpath request arrives, the source node attempts to establish a lightpath through each of the routes from the ordered list in sequence, until a route with the required slots is found. If no available route with required slots is found among the list of alternate routes, the lightpath request is blocked. The computation complexity of this algorithm is higher than that of FR. However, it provides lower call blocking compared to the FR algorithm. Using this algorithm, it may not be possible to find all possible routes between a given source-destination pair, and hence the performance of this algorithm is not optimum in terms of call blocking.

Least congested routing (LCR) [6, 80] similar to FAR; it precomputes a sequence of routes for each source-destination pair. Depending on the arrival time of the lightpath request, the least-congested route is chosen from the precomputed routes.

In adaptive routing (AR) [80, 81], the route between the source-destination pair is selected dynamically, based on link information in a network. A lightpath request is considered blocked when no route with a desired slot is found between the source-destination pair. Since it considers all possible routes between a source-destination pair, it provides the best performance in terms of call blocking. However, its operational complexity is the highest among other routing policies.

Exercises

1. Estimate the fragmentation impact for the link mentioned in Fig. 6.7 using the external fragmentation metric, the entropy-based fragmentation metric, and the access blocking probability metric. Note that, for access blocking probability metric, two types of granularities are used by transponders; three and four slots are considered to satisfy the first and second granularities.

Figure 6.7: Optical link; white and gray slots indicate free and occupied slots, respective.

2. Consider a network and spectrum conditions mentioned in Fig. 6.8. Consider six lightpath requests, which are AB, AC, AD, BC, BD, and DC, arrive in the network sequentially, where each lightpath request is indexed by $i \in \{1, \cdots, 6\}$. Compute fragmentation and contiguous-aligned available slot ratio in the network under the following conditions.

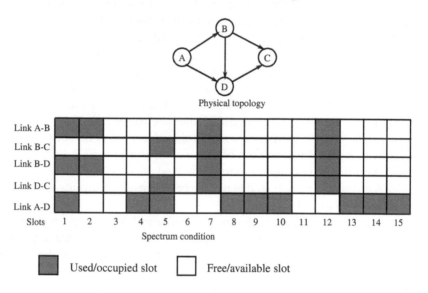

Figure 6.8: Network topology.

   i   The routes of AB, AC, AD, BC, BD, and DC are A-B, A-B-C, A-B-D, B-C, B-D, and D-C, respectively.

   ii The first fit allocation policy is used for spectrum allocation.

   iii Each request requires two contiguous slots for lightpath establishment, and no spectrum conversion is allowed.

   iv No lightpath is tore down after the establishment.

   v   No gardband is considered.

3. If we consider first-last fit on Exercise 2, does it affect the performance in terms of fragmentation and contiguous-aligned available slot ratio? Provide the analysis. For the first-last fit policy, odd and even indexed lightpath requests are allocated using first fit and last fit, respectively.

Chapter 7

Spectrum Fragmentation Management Approaches Considering Non-defragmentation

Spectrum fragmentation management approaches are used to handle spectrum fragmentation and increase the admissible traffic levels in the network. Figure 7.1 shows the thematic taxonomy of fragmentation management approaches. Note that non-defragmentation and defragmentation approaches are not mutually exclusive. One can make some schemes considering both non-defragmentation and defragmentation approaches. This chapter presents spectrum fragmentation management approaches considering non-defragmentation. Spectrum fragmentation management approaches considering defragmentation approaches will be explained in Chapter 8.

7.1 Non-defragmentation approaches

In the non-defragmentation approach, necessary precautions are taken to avoid fragmentation before the establishment of a lightpath. However, no action is taken for in-service lightpaths. Whereas in the case of defragmentation approaches, a necessary action is taken for in-service lightpaths in order to suppress the fragmentation effect. In the non-defragmentation approach, the spectrum is managed in advance to avoid the fragmentation effect. The non-defragmentation

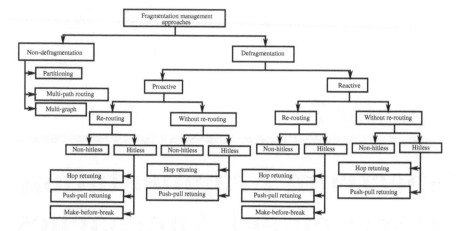

Figure 7.1: Classification of fragmentation management approaches in EONs.

approaches [88, 102, 163–171] are attractive as they offer lower capital expenditure (CAPEX) and operational expenditure (OPEX). However, these approaches have a lower performance in terms of admissible traffic volume than the defragmentation approaches. In the non-defragmentation approaches, the following strategies are used to suppress the spectrum fragmentation.

7.1.1 Partitioning approaches

In EONs, typically there are two types of partitioning approaches, namely dedicated partitioning and pseudo partitioning, are considered, which are explained in the following.

7.1.1.1 Dedicated partitioning

Dedicated partitioning approaches [100, 114, 162] are the more strict version of the pseudo partitioning approach. In dedicated partitioning approaches, the entire spectrum is divided into a number of partitions based on certain criteria, and each partition is dedicated to a different lightpath group. In the following, the criteria to obtain the number of partitions is explained.

In dedicated partitioning approaches, the route and slot demand are assumed to be given for each lightpath group [100, 162]. The number of required partitions can be estimated using the graph coloring problem [85]. The approach uses an auxiliary graph where the lightpath groups are considered as nodes and, if two lightpath groups share at least one common link, they are connected by an edge. A lightpath group is formed by a set of lightpaths whose routes are exactly the same. The graph coloring problem assigns a color to each vertex while satisfying the constraint that the same color is not assigned to adjacent vertices; the objective is to minimize the number of colors. Each color corresponds to each partition unit. Minimizing the number of colors is equivalent to minimizing the number of partitions of the entire spectra.

The dedicated partitioning approaches ensure that the number of partitions must be greater than two. The lightpath groups and number of partitions are created in advance; when a lightpath request arrives, the network checks which category the lightpath request belongs to and assigns it to the appropriate partition. Through careful partitioning, spectrum under-utilization which is caused by unfairness can also be eliminated. The main disadvantage of partitioning techniques is that they fail to offer statistical multiplexing gain. Due to the lack of statistical multiplexing gain, blocking probability can increase in the network if the number of partitions is high [100, 162]. For an example, we estimate the blocking probability, denoted by P_b, using Erlang B loss formula, which is mentioned in (7.1) under a simple traffic model with a Poisson arrival process of mean arrival rate λ and an exponential distribution of the mean lightpath holding time (h) [172]. If the number of channels is 100 and offered traffic, denoted by $E = \lambda h$ is 100 [Erlang], the blocking probability is 0.0757. Dividing the same channel resources among four partitions and splitting the traffic among the partitions (25 channels with offered traffic volume of 25 [Erlang]), the blocking probability for each partition becomes 0.1438, which is higher than that of the non-partitioning case.

$$P_b = \frac{\frac{E^m}{m!}}{\sum_{i=0}^{m} \frac{E^i}{i!}},$$

(7.1)

where m is the number of identical parallel channels.

Figure 7.2 shows the spectrum allocation of lightpath requests based on the dedicated partitioning approach. For this purpose, we consider a three-node network, which is shown in Fig. 7.2(a). We assume that all lightpath requests are categorized into six lightpath groups; the routes of the lightpath groups are given in advance as shown in Fig. 7.2(b). The auxiliary graph, see Fig. 7.2(c), is formed using lightpath groups, where each vertex is represented as a lightpath group. If two lightpath groups share one or more common links, an edge is established between the two vertices of the auxiliary graph. Thereafter, we solve the graph coloring problem [85] to determine the number of partitions, see Fig. 7.2(d). Typically, the number of used colors should be equal to the number of partitions. Thus, the entire spectrum is divided into three partitions and lightpath requests are allocated into those partitions according to the same color, which are shown in Fig. 7.2(e). Figure 7.2(f) shows the spectrum allocation of lightpath requests without considering the dedicated partitioning approach.

In the following section, how lightpath groups and partitions are formed are explained in detail. We define a lightpath group as a set of lightpaths whose routes are exactly the same. We use the term of disjoint lightpaths that do not share any link. The term of non-disjoint lightpaths belong to different lightpath

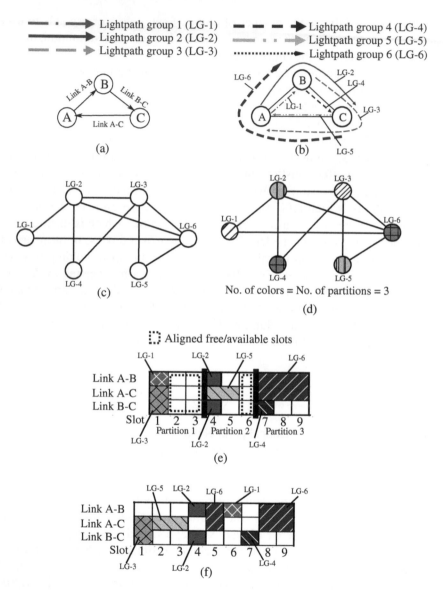

Figure 7.2: (a) Physical network, (b) routes of lightpath groups given in network, (c) auxiliary graph using lightpath groups (d) one of possible solutions of graph coloring problem, (e) spectrum allocation with dedicated partitions, and (f) spectrum allocation without considering partitions.

group and share a link. Our objective in partition allocation is to determine the minimum number of required partitions that accommodate all lightpath groups in the network with the constraint that lightpaths assigned in the same partition must be disjoint. We transform the partition assignment problem into a graph coloring problem by creating a graph, named lightpath group graph. The lightpath group graph indicates the relationship among the lightpath groups in the network.

Algorithm 1 shows the procedure used to create the lightpath group graph. The lightpath group transformation partition is described as follows. The route of each lightpath group is assumed to be given. A vertex of the lightpath group graph corresponds to a lightpath group per unit slot demand. For an example, a lightpath group that has two unit slot demands, yields two vertices. If two lightpath groups share a common link or more links, an edge is established between the two vertices. By default, vertices that correspond to the same lightpath group are connected by edge(s).

Algorithm 1 Lightpath group graph transformation

Input: All lightpath groups in the network

Output: Lightpath group graph

Step 1: Initialize the set of vertices $V = \{\emptyset\}$ and the set of edges $E = \{\emptyset\}$.

Step 2: Vertex generation: Generate vertex v that corresponds to each lightpath group per unit slot demand, where $v = 1, 2, \cdots, |V|$ for $|V|$ paths, and then add v to V.

Step 3: Edge establishment: Establish edge (v, u) between $v \in V$ and $u \in V$ if the two lightpath groups corresponding to vertices v and u share at least one link, add (v, u) to E.

The graph coloring problem assigns a color to each vertex while satisfying the constraint that the same color is not assigned to adjacent vertices. Each color corresponds to each partition unit. A partition unit is a measurement unit indicating partition size. The minimum number of colors means the minimum number of partition units. After the minimum number of partition units is obtained, partition units that belong to the same lightpath group are put in adjacent order and merged into one partition. Hence, the lightpath group that contains larger slot demand is assigned to more partition units, and thus has a larger size partition.

The graph coloring problem can be formulated as an integer linear programming (ILP), which will be discussed in Chapter 11. From the literature [173], it had been revealed that when the number of lightpath groups and/or size of the traffic volume becomes large, the computational complexity of the ILP increases and it becomes difficult to solve it within a practical time. In that situation, the largest degree first (LDF) algorithm [173] can be applied to solve the graph coloring problem. The LDF algorithm attempts to color the vertices in a descending order of degree. LDF is a sequential coloring heuristic that attempts to color vertices on the basis of a specified order by using the minimum indexed color that is not used by adjacent vertices. In sequential ordering, if a vertex receives a particular color once, its color remains unchanged thereafter. The details of the LDF algorithm are presented in Algorithm 2. The time complexity of Algorithm 2 is $O(|V|^2)$, where V is the set of nodes in the network.

Algorithm 2 Largest degree first (LDF)

Step 1: Select the uncolored vertex with the largest degree.

Step 2: Choose the minimum indexed color from the colors that are not used by adjacent vertices.

Step 3: Color the selected vertex using the color described in step 2.

Step 4: If all the vertices are colored, LDF stops. Otherwise, LDF returns to step 1.

Once the partitions are determined, the first-last fit spectrum allocation policy is adopted for spectrum allocation, this has already been explained in Chapter 4 (page number 44). The first-last fit spectrum allocation policy always attempts to choose the lowest indexed slots in the odd number partition from the list of available slots. For the even number partition, it attempts to choose the highest indexed slots from the list of available slots. The first-last fit allocation policy adopted in the partition scheme is expected to provide more contiguous aligned available slots.

The following steps are considered to estimate the overall time complexity of the first-last fit spectrum allocation policy. The time to check the arriving lightpath requests and find appropriate lightpath groups is $O(|Z| \log |Z|)$, where Z is the set of lightpath requests. The time to check the partitions for assigning lightpath groups is $O(|C| \log |C|)$, where C is the set of lightpath groups. The time to perform spectral allocation for Z lightpath requests using a given route is $O(|E||B||Z|)$ or $O(|V|^2|Z|)$. E, B, V are the sets of links in the network, spectrum slots in each link, and nodes in the network, respectively. In the above steps, the third step is the dominating factor. Therefore, the overall time complexity of the spectrum allocation policy is $O(|V|^2|Z|)$.

Dedicated partitioning approaches presented in the literature divide the spectrum into a number of partitions in advance by assuming traffic demands in order to suppress the fragmentation in the network. Considering that the traffic condition can never be fully predicted, it may happen that a lightpath with a large number of required slots may not be supported by a partition; in this situation, a lightpath request tends to be rejected. This will increase call blocking in the network. In addition, dedicated partitioning approaches fail to offer statistical multiplexing gain. Due to the lack of statistical multiplexing gain, blocking probability can increase in the network if the number of partitions is high. To overcome these issues, pseudo partitioning approaches have been considered, which are explained in the following section.

7.1.1.2 Pseudo partitioning

Pseudo partitioning approaches [109, 114, 161, 174] are used to revive unsuccessful lightpath requests by dividing them into two groups and allocating them to different ends of the spectrum. For example, a lightpath request that demands the bandwidth in the Tb/s range is allocated resources from the lower end of spectrum [114]. On the other hand, smaller requests are allocated resources from the higher end of the spectrum [114]. The approach presented in [114] avoids the direct accommodation of mixed types of lightpaths. The criteria for forming lightpath groups depends on the parameters used. Lightpath groups may be estimated based on a higher bandwidth demand or lower bandwidth demand. Lightpath groups may also be created based on disjoint and non-disjoint routes of lightpath requests [161].

Figure 7.3 shows the spectrum allocation of lightpath requests based on the pseudo partitioning approach. In this case, the lightpath requests that demand high bandwidth are allocated to the lower end of spectrum. On the other hand, the lightpath requests that demand low bandwidth are allocated to the higher end of the spectrum.

In the following, we explain how the lightpath groups are created based on disjoint and non-disjoint routes of lightpath requests. To separate the disjoint and non-disjoint lightpaths, lightpath requests are categorized into two groups, namely (i) disjoint lightpath group and (ii) non-disjoint lightpath group. The disjoint lightpath group is the set of lightpaths whose paths are disjoint to each other. The non-disjoint lightpath group contains the remaining lightpaths whose paths do not belong to the disjoint lightpath group. To provide a higher number of aligned available slots, the routing policy that maximizes the number of disjoint lightpaths in the network is adopted.

The disjoint lightpath group can be created using an integer linear programming (ILP) approach, which will be discussed in Chapter 11. When the network size becomes large, the computational complexity of the ILP model increases and it becomes difficult to solve within a practical time. Therefore, a heuristic approach is introduced, which consists of two algorithms for transforming the problem into a graph, and creating the disjoint lightpath group in order to maximize the traffic demand.

In Algorithm 3, multiple paths for all the source-destination pairs are determined in advance. We can adapt any routing policy to determine the multiple paths, such as k shortest paths [85]. We assume that the traffic demand of each

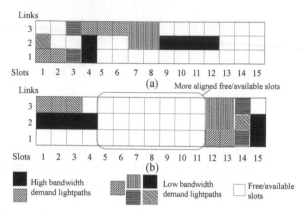

Figure 7.3: Allocation of lightpath requests (a) without pseudo partitioning and (b) with pseudo partitioning.

Algorithm 3 Path graph transformation

Input: All paths for all source-destination pairs

Output: Path graph

Step 1: Initialize the set of vertices $V = \{\emptyset\}$ and the set of edges $E = \{\emptyset\}$.

Step 2: Vertex generation.

> **2.1:** Generate vertex v that corresponds to each path, where $v = 1, 2, \cdots, |V|$ for $|V|$ paths, and then add v to V.
>
> **2.2:** Generate vertex value w_v, which corresponds to each traffic demand of the path associated with vertex v.

Step 3: Edge establishment.

> **3.1:** Establish edge (v, u) between $v \in V$ and $u \in V$ if the two paths corresponding to vertices v and u are multiple paths of the same source-destination pairs, add (v, u) to E.
>
> **3.2:** Establish edge (v, u) between $v \in V$ and $u \in V$ if the two paths corresponding to vertices v and u share at least one link, add (v, u) to E.

---

source-destination pair is known in advance, and one of multiple paths belonging to the same source-destination pair is used with the given traffic demand. After obtaining the set of paths for all source-destination pairs, these paths are transformed into a graph, named path graph. The path graph indicates the relationship among the multiple paths of all the source-destination pairs.

The details of the path graph transformation is presented in Algorithm 3. Step 1 initializes the graph, whereas step 2 generates vertices of the graph. A vertex corresponds to a path. Each vertex is assigned a value that corresponds to the traffic demand of the path. Step 3 generates the edges of the graph. It establishes an edge between two vertices that belong to the multiple paths of the same source-destination pairs. This guarantees that the multiple paths of the same source-destination pairs are not be assigned in the disjoint connection group together. It also establishes an edge between two vertices whose corresponding paths are not disjointed. This ensures that all the members of the disjoint connection group have disjoint paths from each other.

The following steps are considered to estimate the overall time complexity of the path graph transformation. The time to generate vertices is $O(|V|)$. The time to establish edges between vertices that belong to the multiple paths of the same source-destination pairs is $O(|V|^2)$. The time to check the link-disjointness for every pair of paths is $O(|B|^2)$. V and B are the sets of nodes in the network and spectrum slots in each link, respectively. In the above steps, the second step is the dominating factor. Therefore, the overall time complexity of the spectrum allocation policy is $O(|V|^2)$.

After transforming all the paths into the path graph, we maximize the total traffic demands in the disjoint connection group. We introduce a largest value first algorithm to select the appropriate member of the disjoint connection group. The largest value first algorithm is presented in Algorithm 4. This algorithm assigns the member of the disjoint connection group in descending order of vertex value, where each vertex value represents a traffic demand. In the initial stages, all the vertices are set to be unmarked vertices. This algorithm selects an unmarked vertex with the largest value, and marks this vertex. If no adjacent vertex with the selected vertex belongs to the disjoint connection group, the selected one is

---

Algorithm 4 Largest value first

Input: Path graph

Output: Disjoint lightpath group with maximum traffic demand

Step 1: Select the unmarked vertex with the largest value, mark the selected vertex.

Step 2: If no adjacent vertex with the selected vertex belongs to the disjoint connection group, go to step 3. Otherwise, go to step 4.

Step 3: Put the selected vertex into the disjoint connection group.

Step 4: If all the vertices are marked, the algorithm stops. Otherwise, go to step 1.

---

put into the disjoint connection group. This algorithm repeats the same procedure until all the vertices are marked. The time complexity of Algorithm 4 is $O(|V|^2)$.

Once the disjoint and non-disjoint lightpath groups are determined, the first-last-exact fit spectrum allocation policy is adopted for spectrum allocation. The first-last-exact allocation policy is a combination of two allocation policies, namely, (i) first-exact fit and (ii) last-exact fit. The first-exact fit allocation policy is performed on lightpaths whose paths belong to the disjoint lightpath group. Last-exact fit allocation policy is performed on lightpaths whose paths belong to the non-disjoint lightpath group.

7.1.2 Multipath routing and multigraph approaches

In literature, other non-defragmentation approaches, such as Multipath routing and Multigraph approach, are reported. Multipath routing is already explained in Chapter 4.

Multigraph approach [175, 176] was introduced to suppress the bandwidth fragmentation and improve the traffic admissibility. In this approach, $N - b + 1$ graphs are generated, where b is the number of required slots for each request and N is the total number of slots between two nodes. These graphs are produced by considering each edge of a multigraph. Each multigraph is allowed to have multiple edges (also called parallel edges) that have the same end nodes. Thus, two vertices may be connected by more than one edge in a multigraph. The edges of a multigraph are typically mapped onto a single edge of each generated graph whose cost is determined by applying a specific cost function, which considers all b edges. To select the best path, a shortest path algorithm is executed for each generated graph.

7.2 Related works on non-defragmentation approaches in EONs

Non-defragmentation approaches in EONs employ various strategies to enhance network performance and improve the utilization of spectrum resources. This section presents a comprehensive survey of state-of-the-art non-defragmentation approaches for EONs. We analyze the surveyed approaches by elucidating their strengths and weaknesses.

7.2.1 Multipath routing

To suppress the bandwidth fragmentation effect, Zue et al. [177] presented different online service provisioning algorithms, which incorporate dynamic routing, modulation, and spectrum allocation with hybrid single-path routing or multipath routing schemes. They mainly investigated two types of hybrid single-path routing or multi-path routing schemes for online path computation and fixed path sets. They analyzed several path selection policies in terms of suppressing bandwidth fragmentation and maximizing network throughput to optimize the design and evaluated the introduced algorithms using a Poisson traffic model for two mesh network topologies.

The authors in [178] introduced a multi-path fragmentation-aware routing, modulation and spectrum allocation scheme for advance and immediate reservation requests in EONs. To suppress the fragmentation and solve the problem of resource scarcity, the authors introduced a splitting scheme; it splits requests into different parts and transfers these parts onto one or more lightpaths with the use of sliceable bandwidth variable transponders. These provide a mathematical model in order to solve the problem and also present a heuristic algorithm for fragmentation measurement.

To enhance the spectrum utilization and suppress the bandwidth fragmentation in EONs, traffic grooming and multipath routing techniques were presented by Dharmaweera et al. [92]. In their research work, they first investigated the possible gains by jointly applying the two methods with a practical model that considers physical impairments. They assigned spectrum resources and measured the gains in spectral utilization effectiveness over existing approaches. Thereafter, they presented a heuristic for large networks and introduced an analytical optimization formulation for small networks.

7.2.2 Multigraph approach

Moura et al. [175] introduced a multigraph shortest path (MGSP) algorithm in order to suppress the blocking ratio in EONs. The allocation decisions in the MGSP algorithm are based on cost functions, which try to capture the potentiality of spectrum fragments of allocating incoming requests. The authors observed that the MGSP algorithm reduces the bandwidth-blocking ratio four times compared to existing considered RSA algorithms.

In [176], an energy-aware multigraph shortest path routing policy is presented by considering different modulation levels. The numerical results of the presented work indicates that the presented algorithm can save up to 34% energy and suppress bandwidth blocking significantly compared to existing algorithms.

7.2.3 Spectrum partitioning

Fadini et al. [100, 162] introduced a dedicated partition scheme that reduces call dropping by separating the disjoint and non-disjoint lightpaths into different partitions. The works in [100, 162] partition the whole spectrum in advance to manage spectrum resources proficiently with the restriction that lightpaths assigned in the same partition must be link disjoint. The introduced partition approach provides more aligned free slots by extricating the spectrum allocation of non-disjoint lightpaths. More contiguous, aligned available slots between two partitions are created by using the first-last fit allocation policy. As a result, the scheme introduced in [100, 162] reduces call blocking over the network. Their work indicates that, while the number of partitions is minimized, network performance increases. They observe that the dedicated partition scheme has a disadvantage in terms of blocking probability if the number of partitions is large. This is because the statistical multiplexing gain is lost [179].

To overcome the problem of dedicated partitioning, Chatterjee et al. [161] provides a pseudo partitioning scheme that adopts the first-last-exact fit allocation policy for EONs, in order to enhance the number of aligned free slots; small contiguous free slots are ignored. The introduced scheme [161] allocates disjoint and non-disjoint lightpaths separately. Lightpaths with disjoint paths are assigned using the first-exact fit policy. On the other hand, lightpaths with non-disjoint paths are assigned using the last-exact fit policy. This separation of disjoint and non-disjoint lightpaths offers a higher number of aligned free slots. The exact fit policy offers a higher number of contiguous free slots. The contiguous aligned available slots suppress bandwidth fragmentation, and thus the blocking probability is suppressed.

Wang et al. [114] introduced a fragmentation management approach for EONs, which averts bandwidth fragmentation without rerouting lightpaths. If lightpath requests with low and high bandwidths are not distinguished during spectrum allocation, higher-bandwidth lightpath requests may be rejected. To avoid this shortcoming, they investigated an admission control policy that captures the distinct challenges postured by heterogeneous bandwidths. A preventive admission control policy that follows the spectrum partitioning approach is adopted in order to achieve higher provisioning efficiency. It lowers the unfairness of bandwidth assignment and so avoids call blocking. In other work [99], a traffic admission control policy is described that focuses on the special challenges posed by heterogeneous EON bandwidths. They observe that partitioning the spectrum is superior to non-partitioning as it overcomes the difficulties of fragmentation and fairness.

7.2.4 Rerouting

Zhang et al. [103] introduced a bandwidth fragmentation management algorithm that suppresses bandwidth fragmentation in EONs. Their algorithm achieves better resource utilization through proactive network reconfiguration, with the rerouting of established lightpaths. In every phase, the algorithm first chooses some lightpaths, and then decides how to reroute them with the fragmentation management strategy based on routing and spectrum assignment. Finally, the algorithm performs rerouting with best effort traffic migration in order to suppress traffic disruption.

The work in [181] focuses on a disruption-minimized optical path rerouting scheme that alleviates spectrum fragmentation and achieves better spectrum utilization in EONs. The scheme performs minimal rerouting in order to avoid service disruption, since disruption can significantly degrade network services. They then presented a rerouting scheme that establishes alternative routes and allocates frequency slots before releasing the original lightpaths in order to suppress the bandwidth fragmentation in the network.

Singh et al. [169, 170] presented two fragmentation management approaches, namely, (i) a reactive-disruptive scheme and (ii) a proactive-non-disruptive scheme in order to suppress bandwidth fragmentation in EONs. The reactive-disruptive reconfiguration approach allocates incoming lightpath requests, which are unable to serve, by reconfiguring some existing lightpaths. Whereas, the proactive-non-disruptive reconfiguration approach does not disturb existing lightpaths. It allows reconfiguring of only such lightpaths, which can be shifted without crossing over the spectrum of other lightpaths.

Reference [180] presents a traffic accommodation strategy for EONs based on a continuous greenfield grooming approach with spectrum defragmentation, which is similar to storage-disk defragmentation schemes.

A significant number of studies on the different non-defragmentation approaches have been reported in the literature. Table 7.1 summarizes the major works on the different non-defragmentation approaches.

7.3   Performance analysis of non-defragmentation approaches

This section evaluates and analyzes the performances of different non-defragmentation approaches in terms of blocking probability and contiguous-aligned available slot ratio. The contiguous-aligned available slot ratio in the network was defined in Chapter 6 (page number 66). The blocking probability is defined as the ratio of the number of blocked lightpaths to the number of lightpath requests in the network.

7.3.1   *Simulation assumptions and considered parameters*

The following assumptions are commonly used for the purpose of simulation. We adopt NSFNET [6] and the Indian network as the network topology. Each link between two nodes is considered to be bi-directional. The channel spacing and the total number of spectrum slots per channel are taken to be 12.5 GHz and 400, respectively. Lightpaths are generated randomly based on a Poisson distribution process considering arrival rate (λ) and the holding time of lightpath requests follows an exponential distribution with an average value (h). The source-destination pair for each lightpath request are randomly generated. The

Table 7.1: Summary of existing studies on non-defragmentation approaches.

Approaches	Reference	Observations and comments
Multipath routing	Chen et al. [88]	Improves spectral efficiency and reduces blocking under dynamic traffic conditions
	Ruan et al. [89]	Achieves higher spectral efficiency than the existing single-path provisioning scheme
	Ruan et al. [90]	Investigates the network performance metrics, including network throughput and network bandwidth fragmentation
	Zue et al. [177]	Suppresses bandwidth fragmentation and maximizes network throughput
	Zhu et al. [178]	Suppresses blocking probability and improves spectrum utilization compared to existing algorithms
	Dharmaweera et al. [92]	Enhances the spectrum utilization and suppress the bandwidth fragmentation considering both traffic grooming and multipath routing
Multigraph	Moura et al. [175]	Aims to suppress bandwidth fragmentation and increase traffic admissibility
Spectrum partitioning	Fadini et al. [100, 162]	Reduces the call dropping in the network by separating the disjoint and non-disjoint lightpaths into different partitions
	Chatterjee et al. [161]	Enhances the number of aligned free slots and ignores small contiguous free slots by first-last-exact fit allocation policy
	Wang et al. [99, 114]	Shows that partitioning offers better provisioning efficiency than non-partitioning
Rerouting	Zhang et al. [103]	Introduces a bandwidth fragmentation management algorithm in order to suppress bandwidth fragmentation considering rerouting of lightpaths
	Kadohata et al. [180]	Presents a traffic accommodation strategy based on monotonous greenfield grooming approach with spectrum defragmentation

number of required slots for each lightpath request are uniformly distributed as an integer value between 1 to 16, and the average number of requested slots (β) is 8.5. We define the traffic load, in Erlangs, by $v = \lambda \times h \times \beta$. The average slot utilization, denoted by $U(v)$, is estimated as a function of v, which is given by (7.2)

$$U(v) = \frac{v \times \text{Avg. route hops}}{|F| \times |E|}, \tag{7.2}$$

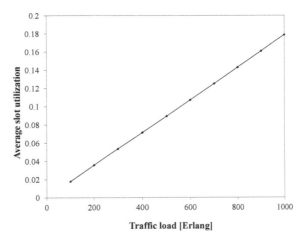

Figure 7.4: Average slot utilization versus traffic load for NSFNET.

where $|F|$ and $|E|$ denote the number of available spectrum slots per link and the number of links in the network, respectively. In our simulation environment, we confirm that NSFNET has the relationship of $U(v) = \frac{v \times 3}{400 \times 42} = 0.0001v$, as shown in Fig. 7.4. All lightpath requests are independent of each other. We ran the simulation using 200 different seeds, for each of which 10000 lightpath requests were generated. The routes between source-destination pairs are estimated using shortest path routing.

For the blocking probability and contiguous-aligned available slot ratio, simulation results are obtained with a 95% confidence interval that is not greater than 5% of the reported average results.

7.3.2 *Numerical results and analysis*

Figures 7.5(a) and 7.5(b) show the blocking probabilities of the dedicated partition and non-partition approaches with different spectrum allocation policies, namely, first fit, random fit, first-last fit for NSFNET and the Indian network, respectively. It is evident from both figures that the dedicated partition approach with first-last fit spectrum allocation policy provides the lowest blocking probability. This is due to the dedicated partitioning approach and the first-last fit spectrum allocation policy. The dedicated partitioning approach provides more aligned available slots and the first-last fit spectrum allocation policy gives more contiguous available slots. As a result, the maximum number of lightpaths is established compared to other schemes. We can also observe that the dedicated partition approach with first fit spectrum allocation policy provides higher blocking probability than that of the non-partition approach with first fit spectrum allocation policy. This is because, the first fit spectrum allocation policy provides more

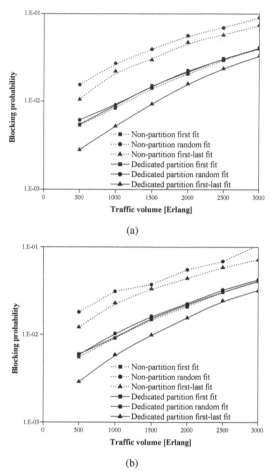

Figure 7.5: Blocking probability versus traffic volume, obtained by using different schemes for (a) NSFNET and (b) Indian network.

contiguous available slots without the partitioning approach. The non-partition approach with first fit spectrum allocation policy has more contiguous available slots than that of the non-partition approach with first-last fit spectrum allocation policy. In the non-partition approach with first-last fit spectrum allocation policy, the number of contiguous available slots is smaller than that of the non-partition approach with first fit spectrum allocation policy as the available slots are squeezed in the middle of the subcarrier slots. Therefore, the blocking probability using the non-partition approach with first-last fit spectrum allocation policy is lower than that of using the non-partition approach with first fit spectrum allocation policy. The random fit spectrum allocation policy provides the worst performance in terms of the blocking probability compared to the other spectrum allocation policies. This is due to the random fit policy allocates slots randomly, and thus the number of contiguous available slots is reduced. The dedicated partition approach with random fit spectrum allocation policy allocates the slots randomly inside each partition and results in more aligned available slots compared to the non-partition approach with random fit spectrum allocation policy. This in turn leads to lower blocking probability compared to the non-partition approach with random fit spectrum allocation policy.

As we already mentioned that partitioning would be beneficial for reducing the blocking probability in elastic optical networks, provided the number of partitions is minimized. Therefore, the performance of the dedicated partition approach is evaluated in terms of the blocking probability as the number of partitions is reduced. The number of partitions in the dedicated partition approach is reduced with the consequence that not all lightpath groups assigned in the same partition are disjointed. Figure 7.6 shows the blocking probabilities of the dedicated partition approach under a different number of partitions and the non-partition approach with first fit spectrum allocation policy for NSFNET. We observe that when the number of partitions is two, the blocking probability is the lowest among other number of partitions. Therefore, we deduce that the reduction of the number of partitions could reduce the blocking probability.

Figure 7.7 plots the blocking probability versus traffic volume obtained by using different non-defragmentation approaches. For comparison, we incorporate the results of dedicated partition approach with two partitions. We observe that the pseudo partitioning approach provides the lowest blocking probability among all non-defragmentation approaches considered. This is because the separation of the disjoint and non-disjoint ligthpaths. This separation provides more aligned available slots thereby reducing the blocking probability in the network. Dedicated partitioning approach also separates the disjoint and non-disjoint ligthpaths. In dedicated partitioning, to avoid statistical multiplexing gain issue, we

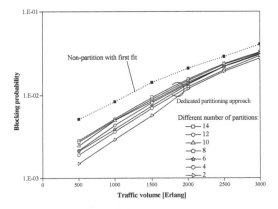

Figure 7.6: Blocking probability versus traffic volume under different number of partitions using partition scheme with first-last fit spectrum allocation for NSFNET.

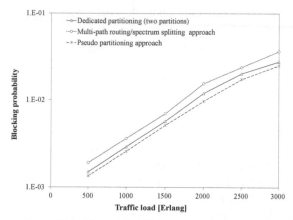

Figure 7.7: Comparison of blocking probabilities using different non-defragmentation approaches.

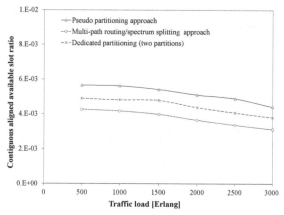

Figure 7.8: Comparison of contiguous-aligned available slot ratios using different non-defragmentation approaches.

reduce the number of partitions by violating the disjoint constraint, and hence it provides more fragmented slots compared to the pseudo partitioning approach. We noticed, but did not include in the figure, that the multipath routing approach yields lower blocking probability than the traditional routing and spectrum allocation approaches. This is because the multipath routing approach splits bandwidth requests into different parts and transfers these parts along one or more lightpaths by utilizing sliceable bandwidth variable transponders. We noticed that the pseudo partitioning approach offers higher contiguous-aligned available slot ratios than the other non-defragmentation approaches considered, which is captured in Fig. 7.8.

In summary, we observed that non-defragmentation approaches suppress the probability of blocking due to bandwidth fragmentation. Among different non-defragmentation approaches considered, the pseudo partitioning approach outperforms the non-defragmentation approaches. As non-defragmentation approaches have lower performance in terms of blocking probability than the defragmentation approaches, the next chapter (Chapter 8) will discuss different defragmentation approaches.

Exercises

1. Why is pseudo partitioning better than dedicated partitioning?

2. Calculate the blocking probability of a fiber link using Erlang B loss formula under the following conditions:

   i Lightpath requests arrive in the system based on a Poisson arrival process and their holding times follow an exponential distribution.

   ii The number of channels is 120 and the offered traffic is 100 Erlang.

3. Estimate the blocking probability of each partition when 120 channels of a fiber link are divided among 10 partitions and splitting the traffic (100 Erlang) among the partitions (i.e., 12 channels with offered traffic volume of 10 Erlang); the traffic assumption remains the same with Exercise 2.

4. Why does the separation of disjoint and non-disjoint requests have better impact in terms of fragmentation in the partitioning approach over without separation of requests?

5. Why is the first-last fit spectrum allocation more suitable for pseudo partitioning in terms of blocking ratio over the first fit policy.

6. Discuss the negative aspects of spectrum split routing.

7. Estimate average slot utilization under the following conditions.

   i Average number of route hops is four.

   ii Traffic load in the network is 100 Erlang.

   iii The network contains 20 links and each link contains 10 slots.

8. What is confidence interval and what does margin of error indicate?

Chapter 8

Spectrum Fragmentation Management Approaches Considering Defragmentation

In Chapter 7, we have already discussed fragmentation management approaches that consider non-defragmentation strategies. This chapter mainly focuses on spectrum fragmentation management approaches that consider defragmentation.

8.1 Defragmentation approaches

The defragmentation approaches are considered to fill-in the gaps left behind after terminating a lightpath. These approaches are typically classified into two main strands: reactive and proactive. Reactive defragmentation approaches are normally triggered when a new lightpath request arrives in the network. On the other hand, proactive defragmentation approaches are applied without waiting for a new lightpath request. Both proactive and reactive defragmentation approaches are again classified into two types, namely with and without rerouting of existing lightpaths. The rerouting approaches [67, 103] reallocate existing lightpaths to the same or different spectrum slots by changing their routes in order to avoid the fragmentation effect. On the other hand, without rerouting, approaches do not allow existing lightpaths to change their routes; spectrum reallocation may be allowed. Based on traffic disruption, both with and without rerouting of existing lightpaths are categorized into the non-hitless and hitless defragmentation approaches, which are discussed in the following.

8.1.1 Non-hitless defragmentation approaches

The defragmentation approaches that cause traffic disruption are referred as non-hitless defragmentation approaches [67, 103]. These approaches attempt to maximize the size of contiguous blocks of unassigned frequency resources with triggering traffic disturbance. As these approaches always cause traffic disruption, they are not preferred in EONs. To overcome this problem, hitless defragmentation approaches are considered for EONs, as is explained in the following.

8.1.2 Hitless defragmentation approaches

The defragmentation approaches that do not cause any traffic disruption are referred as hitless defragmentation approaches. These techniques [182–185] attempt to maximize the size of contiguous blocks of unassigned frequency resources without triggering any traffic disturbance.

Figure 8.1 shows an example of hitless defragmentation and its different conditions. Initially, lightpaths 1 to 4 are active in Fig. 8.1(a). Then lightpath 2 is terminated in Fig. 8.1(b), and we apply hitless defragmentation to retune lightpaths 3 and 4 in Fig. 8.1(c). Finally, lightpath 5 is added to the network in Fig. 8.1(d). In this example, lightpaths are retuned based on proper reconfiguration of allocated spectrum resources. The retuning is executed gradually and the spectrum jumped is not considered. The retuning operations are performed by all involved devices, including filters in intermediate nodes, in a coordinated manner under a distributed control environment or a centralized network controller. Therefore, changing lightpaths from one set of frequency slots to another does not cause any traffic interruption.

To achieve hitless defragmentation, a flexible optical node architecture [184] is essential. Figure 8.2 shows the node architecture that offers hitless defragmentation. It uses a pool of universal transceivers, instead of different types of dedicated transponders (see Fig. 8.3), to satisfy the clients' demand. If dedicated transponders are used, flexibility is insufficient, and hence hitless defragmentation cannot be performed. This is because the synchronization among all involved devices can not be performed in a coordinated manner under distributed control environment or a centralized network controller. In this node architecture for hitless defragmentation, client-side devices no longer include the transceivers; all transceivers are placed in a universal transceiver pool. The client side generates a signal, which is mapped to transport frames, and the modulation format is decided. A bandwidth variable cross-connect switch (BV-WXC), which is placed between the client side and the universal transceivers pool, enables the sharing of transceivers from the universal pool. Using this architecture, the client selects

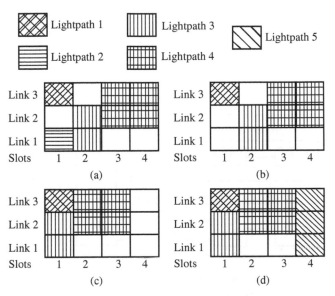

Figure 8.1: Example of different conditions of hitless defragmentation (a) initial state, (b) lightpath 2 terminated, (c) defragmentation using hitless, and (d) lightpath 5 added in network.

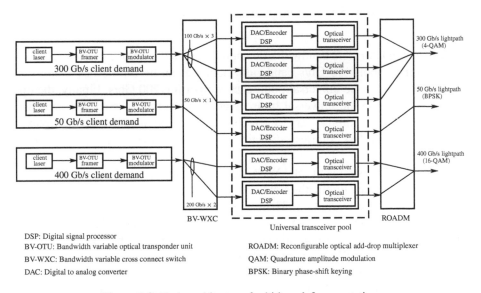

DSP: Digital signal processor
BV-OTU: Bandwidth variable optical transponder unit
BV-WXC: Bandwidth variable cross connect switch
DAC: Digital to analog converter

ROADM: Reconfigurable optical add-drop multiplexer
QAM: Quadrature amplitude modulation
BPSK: Binary phase-shift keying

Figure 8.2: Node architecture for hitless defragmentation.

a suitably configured universal transceiver. For example, to support 300 Gb/s client demand, three transceivers are required, each supporting 100 Gb/s band-

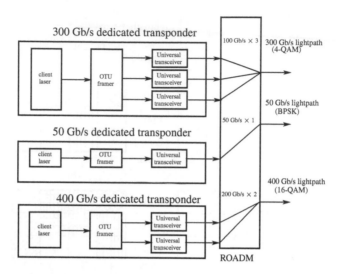

OTU: Optical transponder unit

QAM: Quadrature amplitude modulation

BPSK: Binary phase-shift keying

ROADM: Reconfigurable optical add-drop multiplexer

Figure 8.3: Node architecture using dedicated transponders.

width demand. 50 Gb/s bandwidth demand is satisfied with one transceiver from the universal pool. 400 Gb/s bandwidth demand can be fulfilled by using two transceivers from the universal pool, each supporting 200 Gb/s. Finally, the optical signals from the universal transceiver pool are multiplexed and switched to the appropriate output fibers by a reconfigurable optical add-drop multiplexer (ROADM) that offers colorless, directionless, contentionless, and grid-less properties.

Figure 8.4 illustrates the BV-WXC control during the retuning process in order to move existing lightpaths from one set of frequency slots to another without causing any traffic disruption. There are two types of control, namely sequential BV-WXC control (see Fig. 8.4(a)) and synchronous BV-WXC control (see Fig. 8.4(b)). They are differentiated based on their approaches in the retuning process. Sequential BV-WXC control uses one large step retuning during which the spectrum between the source and destination the bandwidth are not available. On the other hand, synchronous BV-WXC control proceeds retuning by successive small steps and, after each step, available spectrum can be used. Sequential BV-WXC control is simpler and can be used when the retuning time is small compared to the average inter-arrival time of requests. However, if retuning time is not small enough, synchronous BV-WXC control may be preferable.

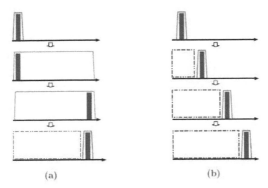

(a)   (b)

Figure 8.4: Illustration of BV-WXC control during the retuning (a) sequential BV-WXC control and (b) synchronous BV-WXC control.

In the literature, there are three main hitless defragmentation or retuning approaches; (i) hop-retuning [182], (ii) push-pull retuning [184], and (iii) Make-before-break. These hitless defragmentation approaches are discussed below.

8.1.2.1 Hop retuning

Hop retuning technology retunes lightpaths to any available spectrum slot regardless of whether it is continuous or not. The technology for hop retuning was introduced by Proietti et al. [182].

Figure 8.5 explains the concept of hop retuning. We consider a simple three-node network scenario for demonstration of hop retuning as shown in Fig. 8.5(a). Figure 8.5(b) shows the spectrum reallocation of the different lightpaths, light-path 1, lightpath 2, and lightpath 3, on links A-B and B-C before and after the defragmentation processes. We assume that each lightpath needs one slot and the central frequencies of slots 1, 2 and 3 are f_1, f_2, and f_3, respectively (see Fig. 8.5(c)). Lightpath request 4 arrives for lightpath establishment from node A to node C. On link A-B, slot 1 is available but the same spectrum slot is not available on link B-C. An efficient usage of the spectrum resource is to move lightpath 1 from slot 1 to slot 3. In this way, spectrum resource can be available on link B-C to establish new lightpath 4. Note that the spectrum reallocation can also be performed when a lightpath occupies few contiguous spectrum slots. This retuning process does not disrupt any existing lightpath and it can be implemented with the help of following technology.

Figure 8.6 shows how to retune existing lightpaths without any traffic disruption. Rapidly-tunable lasers at the transmitter and burst-mode coherent receivers with fast wavelength tracking at the receiver are used for hitless defragmentation. The fast auto-tracing method involves an athermal arrayed waveguide grating (AWG) with a detector array that senses a change in transmission wavelength. The incoming signal at the receiver side is dropped and directed to the input

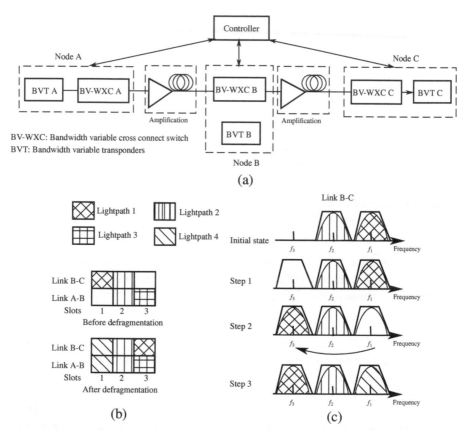

Figure 8.5: Concept of hop retuning (a) three-node network scenario, (b) spectrum realloca-tion before and after defragmentation, (c) defragmentation steps with BV-WXC reconfigura-tion to accommodate new lightpath.

of an AWG with an assured spectrum frequency. The outputs of the AWG then connect to a low-speed photodetector array in order to observe the optical signal power at the various spectrum locations to decide whether the wavelength of the incoming lightpath has changed or not. If the wavelength has changed, the pho-todetector array output triggers a transition from zero voltage to a certain voltage. A field-programmable gate array is required to determine the new wavelength. The rapidly-tunable local oscillators in the transponders are used to track the new wavelength of the incoming lightpath. When the network controller deter-mines to defragment spectrum slots and moves the lightpath from one frequency to another, it first sends control signals in order to configure the corresponding network elements of the intermediate links. The network controller then sends a control message to intermediate nodes to perform defragmentation.

As the hop retuning technology demands sensitive spectrum sensing, it is difficult to implement in finely granular systems. As each rapidly-tunable

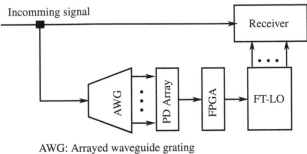

AWG: Arrayed waveguide grating
PD Array: Photodetectors array
FPGA: Field-programmable gate array
FT-LO: Fast tuning local oscillator

Figure 8.6: Illustration of wavelength tracking at coherent receiver using AWG, FPGA, PD array, and FT-LO.

laser/coherent receiver pair covers only a limited range of spectrum slots, the number of photodetectors needed is equal to the number of spectrum slots. Note that for a 12.5 GHz grid system, an AWG with 400 ports and 400 photodetectors is necessary, which increases system complexity. Therefore, the hop retuning approach is unsuitable for finely-granular systems.

8.1.2.2 Push-pull retuning

The push and pull technology [184,185] can cover the entire spectrum grid range. It is executed gradually step-by-step and spectrum jumps are not allowed. It is executed by synchronizing devices under a centralized or distributed control environment. Execution of this technique does not involve re-routing of lightpaths so traffic disruption does not occur.

The time taken for retuning in the push pull approaches is determined by the retuning step width, such as 2.5 GHz, and sweep rate, such as 1, 10, 100, or 1000 ms/step [184]. As an example, in an EON with 12.5 GHz granularity, each step requires five ms per slot, and sweep rate of one ms/step. Similarly, 0.5 sec is required for 100 ms/step. The retuning time for a lightpath is estimated by (8.1)

$$t_{\text{retuning}} = \alpha \times s + \beta, \tag{8.1}$$

where s is the distance between the spectrum index of initial wavelength and the spectrum index of new wavelength. α and β represent retuning time per retuning step and the overall operational time for synchronization, respectively.

Hitless defragmentation is achieved by push-pull retuning, as shown in Fig. 8.7. To avoid traffic disruption, existing lightpaths are continuously swept. The sweep time is one of the significant system parameters for handling dynamic traffic. During a continuous sweep, the compensation of frequency offset between the tunable transmitter laser and the oscillator laser of the receiver

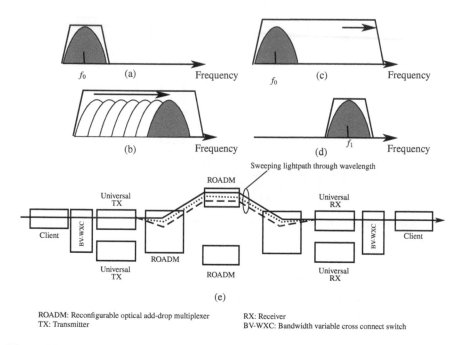

Figure 8.7: Illustration of push-pull retuning (a) initial condition, (b) adjusting for sweeping of established lightpath, (c) continues sweeping of established lightpath, (d) final condition after sweeping, and (e) Sweeping lightpath on the same route through wavelength.

is compulsory in the universal transceiver pool. The total defragmentation time mainly depends on (i) propagation delay between transmitter and receiver, (ii) signaling method for wavelength sweep, and (iii) sweep speed of tunable lasers.

8.1.2.3   Make-before-break technique

The make-before-break technique [181] can also be used to achieve hitless defragmentation. In the make-before-break technique, an additional lightpath between the same source-destination pair is setup while the original lightpath remains active. The routes of original and additional lightpaths should be link disjoint, and the traffic is switched between the two lightpaths. Finally, the original lightpath is torn down, and the traffic carried through the newly-established lightpath. The concept of the make-before-break technique is explained in Fig. 8.8.

The main limitation of the make-before-break technique is that it needs the additional resources and transponders that allow hitless defragmentation. A further issue with the make-before-break technique is the additional operations required in the optical layer. The establishment and release of lightpaths depend on the number of associated fiber links, including optical amplifiers. This forces

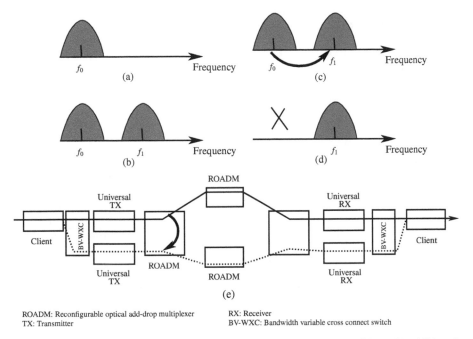

Figure 8.8: Illustration of make-before-break technique (a) initial condition, (b) additional lightpath establishment, (c) traffic switching between original lightpath and additional lightpath, (d) termination of original lightpath, and (e) switching lightpath from original path to new path.

composite optical power equalization, which may affect the stability of other active lightpaths.

8.1.2.4 Comparison of defragmentation approaches

Table 8.1 compares non-hitless defragmentation and hitless defragmentation approaches. The main advantage of non-hitless defragmentation approaches is that they can be deployed without additional equipment. However, they often cause traffic disruption. They are inferior to hitless defragmentation approaches in terms of spectrum defragmentation. As a result, when sensitive data transmission is required without traffic disruption, hitless defragmentation approaches are preferred. Among different hitless defragmentation approaches, hop retuning provides the best performance in terms of suppressing the fragmentation effect. However, hop retuning is not implemented in finely-granular systems, such as 2.5 GHz frequency spacing, due to the current lack of filtering equipment, and equipment costs are high. The make-before-break technique suppresses fragmentation by sacrificing additional resources and transponders, which support retuning in the hitless manner. Push-pull retuning can be used for all levels of spectrum granularity including finely-granular systems, and significantly suppresses bandwidth fragmentation. The defragmentation channel per step indicates the number

Table 8.1: Comparison of different defragmentation approaches.

Evaluation parameters	Non-hitless defragmentation		Hitless defragmentation		
	Rerouting	Without re-routing	Make-before-break	Push-pull re-tuning	Hop retuning
Extra equipment	No	No	Yes	Yes	Yes
Interrupt traffic	Yes	No	No	No	No
Defragmentation channel per step	N/A	N/A	Single channel	Single channel	Multiple channels
Defragmentation speed	N/A	N/A	Fast	Slow	Rapid ($<$ $1\mu s$)
Cost	Low	Low	Maximum	Moderate	High
Complexity	Higher than without rerouting	lowest	Higher than push-pull	Higher than non-hitless	Highest
Defragmentation spectral area	N/A	N/A	Any	Limitation	Any

of channels considered for defragmentation process in each step. The defragmentation channels per step for make-before-break, push-pull retuning and hop retuning are single, single and multiple, respectively.

The different types of defragmentation approaches addressed in the literature are summarized in Table 8.2.

8.2 Related works on defragmentation approaches in EONs

Fragmentation management approaches in EONs employ various strategies to enhance network performance and improve the utilization of spectrum resources. This section presents a comprehensive survey of state-of-the-art fragmentation management approaches for EONs. We analyze the surveyed approaches by elucidating their strengths and weaknesses.

8.2.1 Hop retuning

Zhang et al. [187] introduced two hitless defragmentation algorithms, namely maximum spectrum rejoin (MSR) and minimum number of operations (MNO), in order to maximize spectrum rejoins and to reduce the number of operations. The MSR algorithm is applied in both hop retuning and push and pull approaches, while the MNO is only applied for hop tuning. Their results indicate that both algorithms reduce bandwidth fragmentation in EONs.

Table 8.2: Summaries of different defragmentation approaches.

	Approaches	Non-hitless defragmentation		Hitless defragmentation		
		Continuity constraint	Contiguity constraint	Hop retuning	Push pull retuning	Make-before-break
Proactive	Rerouting	Zang et al. [68] Zang et al. [69]				Takagi et al. [181]
Proactive	No rerouting		Wang et al. [114]	Proietti et al. [182] Zhang et al. [187]	Sekiya et al. [186] Aoki et al. [184] Wang et al. [183] Wang et al. [188]	
		Shi et al. [189], Yin et al. [102] Chen et al. [88], Fadini et al. [100] Singh et al. [169], Zhang et al. [171]				
Reactive	Rerouting	Zang et al. [69] Singh et al. [169]		Zang et al. [69]		
Reactive	No rerouting				Wang et al. [188]	

The authors in [69] investigated the operational principles of hitless defragmentation techniques, and analyzed four questions, (i) how to reconfigure?, (ii) how to migrate traffic?, (iii) when to reconfigure?, and (iv) what to reconfigure?. They introduced a dependency graph based defragmentation scheme in order to suppress blocking probability in the network.

Reference [188] detailed a traffic-disturbance-free defragmentation technology, hitless optical path shift (HOPS); it permits existing setup optical lightpaths to move over available and contiguous spectral bands in a hitless manner. This technology does not alter the end-to-end route of optical lightpaths and does not interrupt other existing established optical lightpaths. Numerical analyses of the introduced HOPS algorithm indicate that hop tuning achieves better defragmentation performance than other hitless defragmentation approaches.

8.2.2 Push-pull retuning

In [188], a hitless optical path shift approach using push and pull retuning is introduced in order to suppress bandwidth blocking in EONs. Initially, the authors consider proactive defragmentation, which maintains the spectrum in a delta state where no lightpath is reallocated. Their introduced algorithm is triggered when a lightpath leaves. Considering that frequent defragmentation upon each lightpath termination is undesirable, they introduce reactive defragmentation. The reactive approach executes defragmentation only when it is necessary or when a lightpath cannot be established without defragmentation. Also, only relevant lightpaths are defragmented to accommodate the incoming lightpath request to avoid lightpath provisioning delay.

Paolucci et al. [190] presented an active stateful path computation element enabling elastic operations and a hitless defragmentation scheme in order to suppress bandwidth blocking in EONs. An online reoptimization algorithm is employed to attempt to move existing lightpaths to allow elastic operations. Finally, the authors address the effectiveness of the introduced algorithm in terms of overall network utilization.

The works in [191, 192] focus on optical-amplified exposing systems, and experimentally validate a push-pull defragmentation technique. The push-pull defragmentation technique is based on dynamic lightpath retuning upon proper reconfiguration of allocated spectrum slots. This technique does not involve extra transponders and does not trigger any traffic disruption in the network. All the relevant technological restrictions related to the push-pull technique are also addressed. An effective closed-form expression is further presented and experimentally demonstrated in order to assess the extreme retuning range in each push-pull operation, which guarantees transmission quality in the network. Thereafter, the introduced technique is demonstrated in a EON in order to evaluate the effect of the introduced approach.

Rozental et al. [193] presented and experimentally validated a synchronous switching technique for push-pull defragmentation based on synchronized transmitter-side and receiver-side interpolators. They show that their progressively executed rate adjustment method with small discrete steps permits a dynamic equalizer at the receiver-side to successfully track the signal changes. The introduced technique may be useful, for example, in reducing power consumption during night-time. Furthermore, the released spectrum may be used to accommodate other short-life opportunistic demands, such as in data center traffic in order to suppress fragmentation effects in the network.

Reference [174, 194] is dedicated to hitless defragmentation for EONs by using the route partitioning approach in order to increase the admissible traffic in the network. The presented scheme partitions the entire routes for all source-destination pairs into two partitions to avoid the interference among lightpaths during retuning. To allocate spectrum, the first-last fit allocation policy is used; one partition is allocated using first fit and the other partition is allocated using last fit. Lightpaths that are allocated on different partitions cannot interfere with each other, and hence the introduced route partitioning scheme prevents interference among lightpaths during retuning. The route partitioning problem is defined as an optimization problem of minimizing the total interference, and they present a heuristic algorithm for large networks, where the existing integer linear programming formulation is not tractable.

8.2.3 Experimental demonstrations regarding defragmentation approaches

This subsection highlights the research works that focus on the experimental demonstrations that have been carried out for spectrum defragmentation in EONs.

Ma et al. [207] experimentally demonstrated a control-plane framework that realizes online spectrum defragmentation in SD-EONs to maximize the admissible traffic and suppress call dropping. Zhang et al. [202] demonstrated SDN over EONs for data center service migration.

An experimental demonstration of an architecture of EONs was conducted by Kozicki et al. [195]. They experimentally setup elastic optical paths and the spectrally-efficient transmission of many channels with data rates varying from 40 to 140 Gb/s among six nodes in a mesh network.

Geisler et al. [206] also experimentally demonstrated a flexible-bandwidth network with a real-time adaptive control plane. They alter the modulation format in order to maintain the required QoT and bit error rate even for signals. An elastic optical network node with defragmentation functionality and novel SDN-based control scheme was experimentally demonstrated, which validates the overall feasibility of extended OpenFlow messages and signal performance [208].

In conclusion, a noteworthy number of experimental demonstrations have been conducted as seen in the literature. Key experimental demonstrations related to the defragmentation process in EONs are summarized in Table 8.3. While the practical deployment of the EON remains under development, fully commercial EONs are expected to become available in the near future.

A significant number of studies on the different defragmentation approaches have been reported in the literature. Table 8.4 summarizes the major works on the different defragmentation approaches.

Table 8.3: Summaries of different experimental demonstrations on defragmentation approaches.

Experimental demonstrations	Reference
Data plane	[35, 195, 196]
Control/management plane	[197–202]
Defragmentation	[203–205]

Table 8.4: Summary of existing studies on defragmentation approaches.

Approaches	Reference	Observations and comments
Hop retuning	Zhang et al. [187]	Introduces two hitless defragmentation algorithms to maximize spectrum rejoins and reduce the number of operations
	Zhang et al. [69]	Investigates the operational principles of hitless defragmentation and suppresses blocking probability
	Wang et al. [188]	Aims for traffic disturbance-free defragmentation to enhance spectrum utilization
Push-pull retuning	Rozental et al. [193]	Validates experimentally a synchronous switching technique based on synchronized transmitter-side and receive-side interpolators
	Paolucci et al. [190]	Introduces an active stateful path computation element enabling elastic operations to suppress the bandwidth blocking
	Sekiya et al. [186]	Considers spectrum retuning on working lightpaths without considering backup paths
	Cugini et al. [191, 192]	Focuses on optical-amplified implementing systems and experimentally validates a push-pull defragmentation technique
	Ba et al. [174, 194]	Proposes hitless defragmentation scheme using the route partitioning approach to increase the admissible traffic
Make-before-break	Takagi et al. [181]	Aims for a disruption minimized make-before-break technique to perform defragmentation
Experimental Demonstrations	Ma et al. [204]	Demonstrates the online spectrum defragmentation using OpenFlow switches in EONs
	Zhang et al. [202]	Demonstrates SDN over EONs for data center service migration
	Kozicki et al. [195]	Demonstrates the architecture of EONs
	Geisler et al. [206]	Experimentally establishes elastic optical paths for EONs

8.3 Performance analysis of defragmentation approaches

This section evaluates and analyzes the performances of different defragmentation approaches in terms of blocking probability. The blocking probability is defined as the ratio of the number of blocked lightpaths to the number of lightpath requests in the network. We consider the same assumptions as that of Chapter 7 (page number 85).

For the blocking probability, simulation results are obtained with a 95% confidence interval that is not greater than 5% of the reported average results.

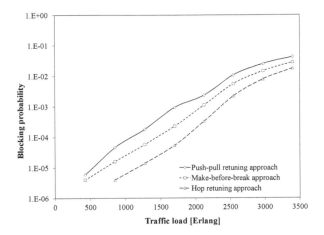

Figure 8.9: Comparison of blocking probabilities using different hitless defragmentation approaches when $\alpha = 0.01$.

In the following, we compare the results of different hitless defragmentation approaches, which are hop retuning, push-pull retuning, and make-before-break approaches.

Figure 8.9 plots the blocking probability versus traffic volume obtained by using different hitless defragmentation approaches. We observe that the push-pull retuning approach yields higher blocking probability than the other hitless defragmentation approaches as it suffers a bottleneck due to end-of-line situations. An end-of-line situation occurs when a lightpath cannot be retuned to fill in a gap left by an expired lightpath due to the interference of another lightpath which prevents it from being moved further. The make-before-break approach offers lower blocking probability than the push-pull retuning approach. In case of the make-before-break approach, lightpath requests are not blocked due to end-of-line situations; the make-before-break approach establishes an additional lightpath between the same source-destination pair along with the original lightpath, and the traffic is continued through the newly-established lightpath. Note that the original lightpath is torn down after establishing the additional lightpath. As a result, the blocking probability is suppressed. The hop retuning approach achieves the lowest blocking probability among all defragmentation approaches considered. This is because it retunes lightpaths to any available spectrum slot regardless of whether it is continuous or not; the bottleneck due to end-of-line situations can be overcome by using the hop retuning approach. Additionally, the hop retuning approach does not require any additional lightpath for bandwidth defragmentation, unlike the make-before-break approach.

For the push-pull retuning approach, the time required to retune a lightpath from an initial position to a new position mainly depends on two parameters,

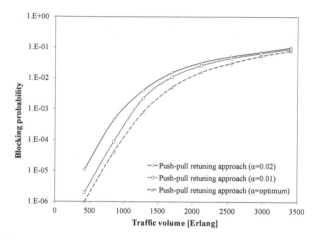

Figure 8.10: Blocking probability versus traffic volume obtained by push-pull retuning approach with different α values.

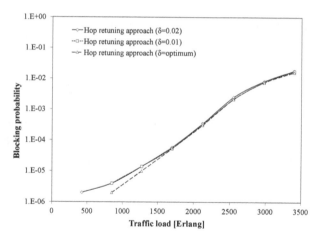

Figure 8.11: Blocking probability versus traffic volume obtained by using hop retuning approach with different δ values.

which are retuning time per retuning step (α) and operational time (β) for synchronization. The blocking probabilities using the push-pull retuning approach for different α values are captured in Fig. 8.10. We observe that, as α increases, the blocking probability increases. This is because several requests arrive in the network before the first defragmentation process is completed. In other words, we can say that the blocking probability is improved with higher retuning speeds.

The blocking probabilities using the hop retuning approach with different retuning speeds are captured in Fig. 8.11. For hop retuning, we assume that it takes δ [time unit] to retune a lightpath. We observe that the blocking probabilities using the hop retuning approach are comparable regardless of the retuning speed. In other words, we can say that the hop retuning approach is not as dependent to the retuning speed as the push-pull retuning approach. This is because the hop retuning approach moves lightpaths from the initial position to the intended position in a single step, while the push-pull retuning approach requires several steps.

In summary, we observe that defragmentation approaches significantly suppress the probability of blocking due to bandwidth fragmentation. Among different defragmentation approaches considered, the hop retuning approach provides the lowest blocking probability.

Exercises

1. Why is the performance of defragmentation approaches better than non-defragmentation approaches in terms of fragmentation issues?

2. Why do hop retuning approaches perform better than other defragmentation approaches in terms of fragmentation issues?

3. What is an end-of-line condition? Explain its negative aspects.

4. The distance between the spectrum index of initial position and the spectrum index of new position of a lightpath is 10.5 nm. The retuning time per retuning step and the overall operational time for synchronization are 1 ms and 10 ms, respectively. How much time is required to perform push-pull retuning for the lightpath. Note that each step covers 1.5 nm distance during retuning operations.

5. What is the difference between proactive and reactive defragmentation approaches?

6. How is make-before-break different from hop retuning and push-pull retuning approaches?

7. What are the disadvantages of hop retuning approaches?

8. Consider the topology mentioned in Fig. 8.12. Assume that each link has five spectrum slots. Initially, all slots for each link are available. Consider 15 lightpath requests, AB, AC AD, AE, AF, BC, BD, BE, BF, CD, EC, FC, ED, FD, and FE, arrive in the network sequentially. Perform spectrum allocation under the following conditions. After lightpath establishment, lightpaths AC and FE will be tore down and then push-pull retuning operations will be performed. Determine contiguous-aligned available slot ratio in the network (i) after tearing down of lightpath requests, but before push-pull retuning and (ii) after push-pull retuning.

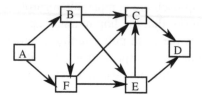

Figure 8.12: Network topology.

i Consider the minimum hop routing and the first fit spectrum allocation policy.

ii Each request requires two contiguous slots for lightpath establishment, and no spectrum conversion is allowed.

iii No gardband is considered.

Chapter 9

Spectrum Fragmentation Management Approaches in 1+1 Protected Elastic Optical Networks

Elastic optical networks (EONs) carry highly reliable traffic and failure of any component or any interruption of traffic flow in network causes massive loss of data and revenue. Therefore, the survivability against the failure has become a crucial requirement for EONs. 1+1 protection is considered as one of the most reliable data transfer techniques in survival EONs, where suppressing spectrum fragmentation is always challenging. This chapter presents and analyzes defragmentation schemes in 1+1 path protected EONs to suppress the call blocking in the network.

9.1 Overview

Several techniques, namely restoration [144], p-Cycles [209], and protection [210], have been considered for survivability purposes in EONs [136, 211]. As the backup resources are reserved by a protection technique prior to fault occurrence, it assures a prompter recovery than other techniques do. Thus, to design a faster recovery system, the protection technique is more preferable.

The protection techniques are typically classified into shared and dedicated protection techniques. Shared protection techniques enhance the resource utilization efficiency but they cannot cover multiple link failures completely. To provide instantaneous recovery and support reliability against multiple link failures, 1+1 protection techniques are considered in survival EONs, where suppressing bandwidth fragmentation to enhance spectrum utilization is always challenging. Chapter 8 has already confirmed that the performance of defragmentation approaches in terms of suppressing fragmentation and call blocking is better than non-defragmentation approaches. Therefore, a defragmentation approach [212, 213] is considered here to handle fragmentation problems in 1+1 protected EONs, which is explained below.

9.2 Demonstration of defragmentation scheme using path exchanging

The presented defragmentation scheme is intended to offer increased traffic load in resilient EONs. For network resiliency, a 1+1 path protection scenario is considered to offer protection against link failures. With 1+1 path protection, each established signal is duplicated and both signals are transmitted to the destination through disjoint paths. This allows the receiver to select the incoming data from any of the two signals. Thus, if one path suffers link failure or is disconnected, the data reception is continued through the other path.

The presented scheme considers that both paths of the 1+1 path protection can be alternately primary and backup paths. In order to perform spectrum defragmentation on primary paths, we simultaneously toggle them and their respective protection paths from primary to backup paths and from backup to primary paths respectively. Toggling a primary path to become a backup path changes its function from being the primary path through which the data is transmitted to become the backup path on standby. Thus, we exchange the function of the primary path to its backup path and vice versa. We allow backup paths on standby to be reallocated, for defragmentation, while the data is being transmitted through the primary path. We suppose that the period of release during the reallocation process is short enough to guarantee the 1+1 protection at almost every moment.

The presented scheme is able to achieve hitless defragmentation on 1+1 path protected networks without requiring any additional equipment. We take advantage of the availability of a by default alternate signal to reallocate lightpaths considering spectrum fragmentation. Since the data is being received through the primary paths, we can afford to reallocate the signals of backup paths during the defragmentation process without disrupting the data transmission. The defragmentation is performed without traffic disruption, provided that no failure occurs on a primary path while its corresponding backup path is being reallocated.

In terms of eliminating spectrum fragmentation, the advantage of the presented scheme is to be able to reallocate both paths of the 1+1 protection for hitless defragmentation without restriction. With the designated primary and backup paths where data from backup paths are used only if there is some impediment on the corresponding primary paths, only backup paths can be reallocated in a hitless defragmentation [214]. The ability to reallocate both primary and backup paths permits a flexible defragmentation that can be performed thoroughly.

Figure 9.1 illustrates the principle of the presented defragmentation scheme with the function of exchanging the primary and backup paths in 1+1 protection network. Consider the network *ABCD* with four active signals $S1 - S4$. The network and its corresponding spectrum before proceeding to any defragmentation is presented in Fig. 9.1(a). In the illustration examples, the links are considered bidirectional for simplicity. The primary and backup paths of each signal are respectively represented by solid lines and dotted lines, and their corresponding spectrum by plain boxes and hatched boxes. On link *AB*, *S3* and *S4* are primary-path signals and *S2* is a backup-path signal.

Figure 9.1(b) presents the network when the designated primary and backup paths with spectrum retuning is used. After moving backup path *S2*, primary paths *S3* and *S4* are retuned. Then, the backup paths are reallocated using the first fit allocation. We can see that, in this particular example, spectrum retuning does not improve the spectrum fragmentation due to the end-of-line situation preventing *S4* from being retuned over *S3*.

Figures 9.1(b) and 9.1(d) show the defragmentation process with the primary and backup paths exchanging. In Fig. 9.1(c), *S4* signal through link *AC*, which is in the backup state is reallocated without path exchanging operation. Then, in Fig. 9.1(d), *S4* through link *AB* is toggled to the backup state while its corresponding backup path on the 1+1 protection through links *AC* and *BC* becomes the primary path (see the network). While it is in the backup state, the lightpath *S4* is reallocated to remove the spectrum fragmentation on link *AB* (see the spectrum).

9.3 Defragmentation scheme using path exchanging scheme

With the aim to allow maximum traffic load in 1+1 path protected EONs, the presented paths exchanging scheme is applied. It reduces the blocking probability by minimizing the spectrum fragmentation. The way the presented scheme is applied depends on the traffic pattern. With static traffic loads, the spectrum state does not change often overtime, but with dynamic traffic loads, the spectrum is in constant evolution. In the case of static traffic load, spectrum fragmentation can be avoided by network planning, and the optimization problem is used in the rare occasions where changes are needed including the cases of network extensions and reconfigurations. On the other hand, for dynamic traffic loads where lightpaths can be added or removed at any moment, the spectrum has to be defrag-

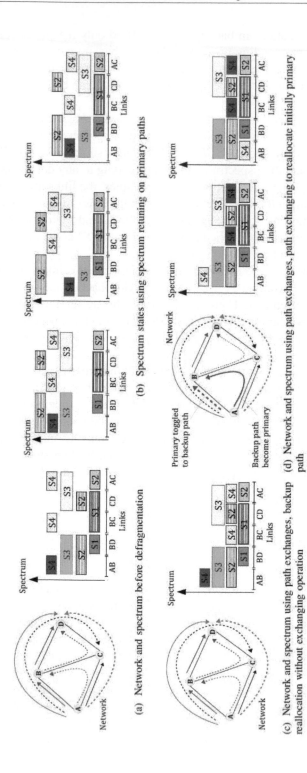

(a) Network and spectrum before defragmentation

(b) Spectrum states using spectrum retuning on primary paths

(c) Network and spectrum using path exchanges, backup reallocation without exchanging operation

(d) Network and spectrum using path exchanges, path exchanging to reallocate initially primary path

Figure 9.1: Example of hitless defragmentation in 1+1 protection. Solid lines and plain boxes represent primary paths, dotted lines and hatched boxes represent backup paths.

mented frequently in order to avoid requests being blocked due to fragmentation. In the following, we focus on dynamic traffic loads.

9.3.1 Dynamic spectrum defragmentation

In our approach to tackle spectrum fragmentation in EONs with dynamic traffic, we suppose that the network can receive a lightpath request at any moment and that allocated lightpaths are active for the requested time periods. The spectrum defragmentation process is triggered at a regular time interval in addition to being automatically triggered whenever a request is blocked. We select to use scheduled defragmentation in a proactive way to prevent blocked requests and limit the processing cost.

It would be preferable to repeat the defragmentation process as much as possible; however that comes with a cost in processing. The scheduled defragmentation offers a compromise as the period between defragmentation processes can be suitably set. Nevertheless, it may happen that between defragmentation processes the cumulative fragmentation causes requests to be blocked. To avoid prolonged blocking until the next scheduled defragmentation, we elect to trigger the defragmentation process whenever a blocking occurs. The overall process of our approach is presented in Fig. 9.2.

9.3.2 Heuristic algorithm

We introduce a heuristic algorithm for the dynamic spectrum defragmentation. The input spectrum is the state at the moment the defragmentation process is triggered. Our introduced heuristic algorithm for dynamic spectrum defragmentation, which is called a mixed primary and backup (MPB) algorithm (see algorithm 5), consists of three steps: (i) sort all lightpaths, both primary and backup paths, in a single list, according to a policy, (ii) proceed to the first defragmentation stage by reallocating lightpaths following the sorted list, and (iii) refine the defragmentation by reallocating lightpaths that can be pushed further down.

In the first step, we sort all lightpaths in a single list, regardless of their initial state (primary or backup) following the selected sorting policy. Since, to reallocate a lightpath in the primary state, we first toggle it to the backup state while its backup is toggled to the primary state, the state of a lightpath may change during the defragmentation process. We consider the path length and the request size as sorting parameters. The length of a path, primary or backup, is the number of link hops on the path. It may be different for corresponding primary and backup paths. A request size represents the number of spectrum slots used by a lightpath. Using these parameters we introduce two sorting policies; the longest path first and the largest slot block first. The slot block area is the product of the path length by the request size.

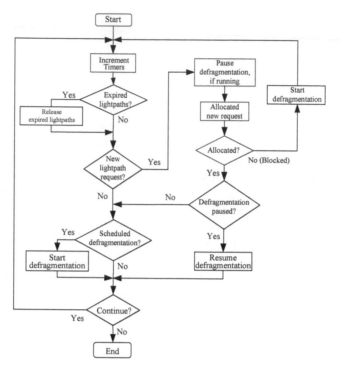

Figure 9.2: Flowchart of dynamic approach.

In the second step, we perform the first stage of defragmentation. Following the sorted list, we reallocate lightpaths to the lower spectrum indexes. At each iteration, we select several lightpaths that can be considered simultaneously. The first selected lightpath is the top of the list. The other lightpaths, which has to satisfy the simultaneous reallocation conditions described as follows, are selected through the list. All the selected lightpaths must be link disjoint and transmitting different signals; we cannot select two lightpaths sharing any link or the primary and backup lightpaths of the same 1+1 protection to avoid reallocation conflict. Once the candidate lightpaths are selected, we reallocate them using the exact-fit policy. The primary lightpaths are toggled to the backup state before reallocation. We refer to the exact-fit allocation policy when we attempt to allocate a lightpath to a free spectrum slot block consisting of exactly the same number of consecutive spectrum slot indexes as the number of spectrum slots required by the lightpath. If there is no such block, we use the first-fit allocation policy. When a lightpath cannot be reallocated to a lower allocation index, it remains at its current allocation index. After that, the selected lightpaths are removed from the list. We repeat the same procedure until all lightpaths are selected.

During the second step of the algorithm, once a lightpath has been selected for reallocation, it is not be revisited. For instance, when a short lightpath previ-

---

Algorithm 5 Mixed primary and backup paths defragmentation.

Input: Spectrum, active signals

Output: Defragmented spectrum

Step 1: Sort all lightpaths according to any policy.

Step 2: First stage defragmentation:

2.1: Select the lightpath at the top of the sorted list.

2.2: Select other lightpaths that can be reallocated in parallel with the first selected lightpath, following the sorted list.

2.3: Reallocate selected lightpaths using the exact-fit policy. If primary path toggle first,

2.4: Remove selected lightpaths from the list.

2.5: Repeat from step 2.1, until all lightpaths are selected.

Step 3: Refining defragmentation:

3.1: Initialize pointer at index 0.

3.2: Increment pointer.

3.3: Reallocate selected lightpaths p using the exact-fit policy. If p is a primary, then first toggle p.

3.4: Repeat from step 3.2, until the highest allocation index is reached.

---

ously allocated to a relatively low spectrum index is reallocated, a longer lightpath, which was selected earlier than the shorter one and was reallocated above the spectrum indexes that were previously used by the short lightpath, is not reconsidered. If the spectrum slots freed from reallocating the short lightpath concatenate with some empty slots, the resulting slot block could be used by the longer lightpath. We take this possibility into account with the third step of the algorithm.

In the third step, we refine the spectrum defragmentation by finding the lightpaths that can still be reallocated to lower spectrum indexes. For that, our approach selects lightpaths to be reallocated starting from the ones allocated at the lowest spectrum index. We initialize a pointer at index 0. Then, after each reallocation iteration, we increment the pointer by 1 and select the lightpaths to be considered for reallocation as the ones allocated at the pointed index. We repeat until the highest allocation index is reached. In other words, we reallocate the lightpaths allocated on lower spectrum indexes first so that the lightpaths allocated on higher spectrum indexes can be reallocated to the vacant spectrum that they left behind. We still use the exact-fit allocation policy in this step and we reallocate a lightpath only to spectrum indexes lower than its currently occupied ones.

9.3.3 Time and spatial complexity

We analyze the time and spatial complexities of the MPB heuristic algorithm. The maximum number of active lightpaths is bounded by the spectrum capacity, which is the total available spectrum slots through all links $|F| \times |E|$. In terms of time complexity, step 1 of the algorithm is performed in $O(|F| \times |E| \times (\log|F| + \log|E|))$, and both steps 2 and 3 are performed in $O(|F|^2 \times |E|)$. In each of steps 2 and 3, each lightpath is reallocated at most once and each reallocation process requires $O(|F|)$, which is the time to find the reallocation spectrum slots. The overall time complexity of the algorithm is $O(|F| \times |E| \times (|F| + \log|E|))$. In networks with large spectrum capacity, where $|F| \gg \log|E|$, the time complexity of the algorithm is $O(|F|^2 \times |E|)$. In comparison, the time complexity of the algorithm used in [214] is also $O(|F|^2 \times |E|)$; the algorithm verifies the spectrum availability before reallocating each backup path.

For the spatial complexity, during each of its steps, the algorithm saves the spectrum state with the information related to the active lightpaths including starting allocation indexes and required slots. From one step to the next, the spectrum state and active lightpaths are updated using the same allocated memory. Since saving both the spectrum state and the active lightpaths requires each $O(|F| \times |E|)$ memory space, the memory requirement of the algorithm is $O(|F| \times |E|)$. The algorithm used in [214] also requires $O(|F| \times |E|)$ space to store the spectrum state and active lightpaths.

9.4 Performance evaluation

This section evaluates and analyzes the performances of the presented defragmentation approach based on path exchanging in terms of blocking probability in the 1+1 protected EONs. The blocking probability is defined as the ratio of the number of blocked lightpaths to the number of lightpath requests in the network. We consider the same assumptions as that of Chapter 7 (page number 85).

For the blocking probability, simulation results are obtained with a 95% confidence interval that is not greater than 5% of the reported average results.

We investigate the impact of the spectrum management approaches in 1+1 protected EONs. We consider three approaches, which are (i) backup-path only reallocation, (ii) path exchanging based defragmentation scheme considering MBP, and (iii) path exchanging based defragmentation scheme considering backup paths first (BPF). The BPF algorithm sorts the lightpaths in the primary state and in the backup state on separate lists. Then, it reallocates all the lightpaths that are in the backup state before considering the list of lightpaths in the primary state. In the backup-path only reallocation approach, named backup-only approach, we reallocate only backup paths, which is the same approach as the hop retuning approach presented in Chapter 8; we do not perform any operation on the primary paths.

Figure 9.3 compares the blocking probabilities obtained by using the three considered defragmentation approaches in a 1+1 protected EON. For comparison purpose, we incorporated the results of non-defragmentation approach. It is indicated that the blocking probability using the non-defragmentation approach is the highest. We observe that the backup-only approach has higher blocking probability compared to both path exchanging based defragmentation schemes. This is because the backup-only approach does not need to perform any action to defragment the fragmented slots caused by primary paths. The blocking probability using the path exchanging based defragmentation scheme considering MBP is lower than that of the path exchanging based defragmentation scheme considering BPF. This is because with the primary lightpaths not moved during the reallocation of the backup lightpaths, the BPF approach has less spectrum flexibility.

Figure 9.4 shows the impact of processing time on blocking probability. We observe that when the processing time decreases, the blocking probability in the network is suppressed.

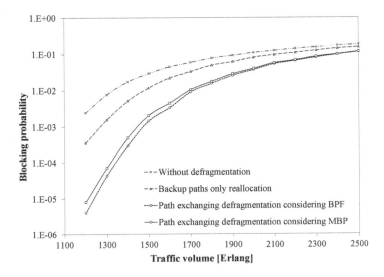

Figure 9.3: Comparison of blocking probabilities.

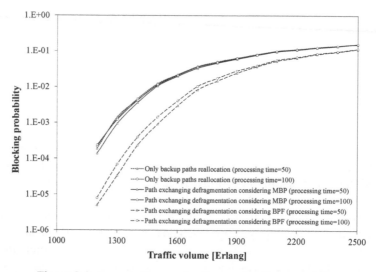

Figure 9.4: Impact of processing time on blocking probability.

9.5 Related works

EONs carry sensitive information and any interruption of the data flow leads to massive data loss. Since optical fibres are subject to impairments, such as being cut, providing network with protection is imperative. This has motivated a continuous research effort to offer protection against failures and provide network resiliency. Several techniques have been considered [211]. They range from, and not exhaustively, span restoration, p-Cycles, 1+1/1:1 end-to-end path protection, and shared backup path protection (SBPP). These techniques were developed in traditional wavelength division multiplexing (WDM) networks and are being extended to EONs. Protection techniques are mainly divided into shared protection and dedicated protection techniques.

Shared protection techniques reduce the spectrum resource used for protection. In [141], the authors considered span restoration in elastic optical networks. They develop integer linear programming (ILP) models to minimize both required spare capacity and maximum used spectrum slots index under different spectrum conversion capabilities. They considered (i) no spectrum conversion, (ii) partial spectrum conversion, and (iii) full spectrum conversion. The authors of [222] considered the p-cycle protection technique, which provides ring-type protection and the speed of restoration of meshes. Similar to [141], they aimed to minimize required spare capacity and the maximum used spectrum slots index and formulated an ILP problem. Both works in [141] and [222] also apply the bandwidth squeezed restoration (BSR) technique [144]. In [223], the authors presented algorithms that provide path protection using p-cycle paths. They claim 100% protection against single and double link failures.

The works in [221, 224] used SBPP in EONs. SBPP proactively reserves backup paths that are independent of the primary paths. When a failure occurs, the signal is recovered from the protection path regardless of where the failure occurs. The authors focus on providing maximally shared capacity on the backup paths for static traffic demand. They present heuristic algorithms in both works. An ILP model is also formulated in [221] for optimal sharing. For dynamic traffic, SBPP in EONs has been presented [225, 226]. The authors of [225] advocate the use of an algorithm applying different strategies for primary and backup resources using first-fit for primary paths and a modified last-fit for backup paths. They aim to reduce the fragmentation and to increase the shareability. In [225], the author evaluated conservative and aggressive backup sharing policies. The sharing policy is considered as aggressive when it only requires the corresponding working lightpaths to be link-disjoint. It is considered as conservative when in addition to the working lightpaths being disjoint, the sharing lightpaths must have the same bandwidth.

Dedicated protection techniques offer resistance against multiple link failures and allow for instantaneous recovery. In [220], dedicated path protection (DPP) has been considered with static traffic demands. The authors focused on the routing and spectrum allocation problem and formulated an evolutionary algorithm to search for optimal solutions. In [214], the authors presented a 1+1 path protection defragmentation approach in dynamic EONs. Their focus is set on the defragmentation advantage offered by the backup paths. They consider that, since the backup paths by nature are used only in case of failure on their corresponding primary path, they can be reallocated and/or rerouted for defragmentation purposes without causing any traffic disruption. Therefore, the authors achieve hitless defragmentation by performing spectrum defragmentation on them.

A survey on the state of the art of survivable EONs is presented in [211]. The authors first review the literature around aspects, such as, spectrum resource sharing among backup lightpaths, and sharing of high-speed optical transponders. Then they discuss the ongoing research issues and future challenges, spectrum defragmentation on path-based protection among others.

A hitless protection switching technique, which offers free switching between working and protection paths without any signal loss, has been introduced in [227, 228]. A similar hitless switching technique has been presented in [229] for protected passive optical network system. The presented scheme applies a hitless path protection switching technique, such as presented in [227–229], to exchange primary and backup paths.

To deploy hitless defragmentation in 1+1 path protected EONs, one can apply the approach presented in [214]. Yet, that may not be enough to eliminate fragmentation since it focuses on defragmenting only the backup paths; the fragmentations caused by primary lightpaths are overlooked. To overcome this issue, and without causing traffic disruption, hitless defragmentation is to be applied on working primary paths too. Researches have been presented for hitless defragmentation on working lightpaths using spectrum retuning [186]. With spectrum retuning, the allocated bandwidth of a working lightpath is swept from its initial position to a new one, while the signal is still being transmitted, to fill in gaps left in the spectrum. Spectrum retuning is therefore a candidate for complementary defragmentation by retuning the primary paths when the approach presented in [214] is used.

Despite applying hitless defragmentation on primary lightpath, we still face the end-of-line phenomenon with spectrum retuning. When spectrum jump is not allowed, the retuning is done gradually and it is subject to the interferences of other lightpaths that share a link with the lightpath being retuned. An end-of-line situation occurs when the retuning process is stopped (if started) due to those interferences. Thus, even when spectrum retuning is applied on primary lightpaths, the scheme presented in [214] is subject to fragmentation caused by primary lightpaths.

To enhance the performance of the defragmentation scheme mentioned in [212], the work in [215, 230] incorporated rerouting of backup paths in toggled-based quasi 1+1 path protected EONs. Therefore, the effectiveness of defragmentation is enhanced in terms of resource utilization.

A significant number of studies on the spectrum management approaches in survival EONs have been reported in the literature, which are summarized in Table 9.1.

Table 9.1: Summary of existing spectrum management approaches in survival EONs.

Reference	Comments
Shen et al. [211]	Study of several survivable techniques in EONs
Wei et al. [141]	Discuss span restoration techniques in EONs
Oliveira et al. [144]	Present path protection using p-cycle paths in EONs
Sawa et al. [215]	Rerouting of backup paths in toggled-based quasi 1+1 path protected EONs
Wang et al. [214]	Present a hitless defragmentation scheme in 1+1 path protected EONs
Hsu et al. [216]	Minimization of spectrum usage for shared backup path protection in EONs
Yadav et al. [217]	Present a multi-backup path protection scheme for survivability in EONs
Oliveira et al. [218]	Protection using failure-independent path protecting p-cycles in EONs
Din et al. [219]	Survivable routing problem with p-cycles protection in EONs
Klinkowski et al. [220]	Discuss dedicated path protection with static traffic in EONs
Walkowiak et al. [221]	Introduce an ILP problem in shared protected EONs
Sekiya et al. [186]	Present a hitless defragmentation on working lightpaths using spectrum retuning in EONs

Exercises

1. Why does the defragmentation scheme using path exchanging provide quasi-1+1 protection?

2. Why does the defragmentation scheme using path exchanging provide better performance in terms of traffic admissibility than push-pull retuning approaches?

3. Explain the necessary conditions to execute the defragmentation scheme using path exchanging operations.

4. Consider a network shown in Fig. 9.5. The spectrum condition of the network is given below. Perform the defragmentation scheme using path ex-

Link	Slot 1	Slot 2	Slot 3	Slot 4
AB	Primary lightpath 1	Primary lightpath 7		
BC	Primary lightpath 2	Primary lightpath 7		
CD	Backup lightpath 3	Primary lightpath 7		
DE	Primary lightpath 4			
EF	Primary lightpath 5			
FA	Primary lightpath 6			
AG	Backup lightpath 1	Backup lightpath 6	Backup lightpath 7	
BG	Backup lightpath 1	Backup lightpath 2		
CG		Backup lightpath 2		Primary lightpath 3
DG		Backup lightpath 4	Backup lightpath 7	Primary lightpath 3
EG	Backup lightpath 5	Backup lightpath 4		
FG	Backup lightpath 5	Backup lightpath 6		

changing, and estimate contiguous-aligned available slot ratio in the network before and after defragmentation. Assume that no lightpath is tore down after the establishment and spectrum continuity and contiguity constraints are maintained.

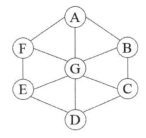

Figure 9.5: Network topology.

5. Consider a network shown in Fig. 9.5. The spectrum condition of the network is given below. Estimate contiguous-aligned available slot ratio in the

Link	Slot 1	Slot 2	Slot 3	Slot 4
AB	Primary lightpath 1	Primary lightpath 7		
BC		Primary lightpath 7		Primary lightpath 2
CD	Backup lightpath 3	Primary lightpath 7		
DE			Primary lightpath 4	
EF		Primary lightpath 5		
FA				Primary lightpath 6
AG	Backup lightpath 1	Backup lightpath 6	Backup lightpath 7	
BG	Backup lightpath 1	Backup lightpath 2		
CG		Backup lightpath 2		Primary lightpath 3
DG		Backup lightpath 4	Backup lightpath 7	Primary lightpath 3
EG	Backup lightpath 5	Backup lightpath 4		
FG	Backup lightpath 5	Backup lightpath 6		

network before and after push-pull retuning. Assume that no lightpath is tore down after the establishment and spectrum continuity and contiguity constraints are maintained.

6. Why does the defragmentation scheme using path exchanging provide better performance in terms of traffic admissibility than only backup path reallocation?

Chapter 10

Spectrum Fragmentation Management in Software-defined Elastic Optical Networks

The flex-grid technology or elastic optical network (EON) is accepted to be a promising solution for the future transport network due to its amazing properties. The software-defined network (SDN) is incorporated with the emerging technology of EONs for enhancing its performance. This chapter exploits spectrum fragmentation management in EONs considering software-defined networks (SDNs).

10.1 Software-defined elastic optical networks

Nowadays, SDN is incorporated with flex-grid technology [231] to enhance resource utilization. With the explosive growth of bandwidth, traditional optical networks are commanding new challenges to future networks. SDN is one of the innovative answers to antiquated solutions. Typically, computing and storage services can utilize the benefits of virtualization and automation. However, due to limited network resources, it is difficult to capitalize on these advantages. SDN has the capability to offer flexibility, control, and a direct path to virtualization with limited network resources.

The software-defined EON (SD-EON) architecture consisting with two separate planes, namely data and control planes, is shown in Fig. 10.1. The data

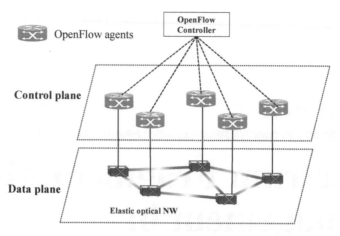

Figure 10.1: Architecture of SD-EON.

plane contains the network elements, such as optical amplifiers, bandwidth-variable transponders (BVTs), and bandwidth-variable cross-connects (BV-WXCs), which transfers data by establishing optical lightpaths. The control plane consists of an OpenFlow controller and several OpenFlow agents. The OpenFlow controller manages the spectrum resources, whereas OpenFlow agents control the operation of data transfers.

The reference model of SD-EONs consisting with three layers, which are infrastructure, control, and application layers, is shown in Fig. 10.2; these layers are stacked over each other. In the data plane, BVTs, optical amplifiers, and BV-WXCs are typically used to construct the infrastructure layer. There are mainly two functionalities performed by these switching devices, which are (i) collecting network information, storing the information temporally into local devices, and directing them to the controllers. (ii) These switches further process the packets according to the guidelines provided by the controller. The control layer acts as a bridge between the application and infrastructure layers through two interfaces, namely south-bound and north-bound interfaces. With the help of the south-bound interface, the control layer initiates controllers in order to utilize the functions of switching devices, which includes the reporting of network status and introducing the rules of packet forwarding. With the help of the north-bound interface, the control layer offers service access points in several shapes, for example, an application programming interface (API). Using this API, SD-EON applications collect network information from different bandwidth variable switching components and build a system based on this information. The application layer is responsible to satisfy users' requirements. SD-EON applications access and manage switching components at the infrastructure layer with the help of the programmable platform delivered by the control layer; SD-EON ap-

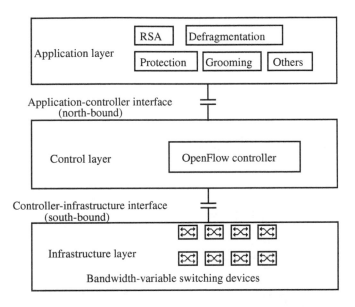

Figure 10.2: Different layers of SD-EON reference model.

plications can include routing and spectrum allocation (RSA), traffic grooming, defragmentation, server load balancing, network virtualization, seamless mobility and migration, and dynamic access control.

The major difference between traditional SDN and SD-EON architectures is due to the network elements that form the data plane. Typically, bandwidth-variable transponders and switching components are used to build the data plane of SD-EON. On the other hand, normal switches are used in the traditional SDN. The resource in the SD-EON control plane is spectrum slots, which is another difference between SD-EON and SDN. Furthermore, when routes are computed, the control plane in SD-EON considers several aspects, such as modulation formats and number of required spectrum slots.

10.2 Architecture of SD-EONs

This section presents the architecture of SD-EONs, which is based on hop-retuning [182]. We already discussed about the hop-retuning in Chapter 8 (page number 97). The architecture on demand [70] is the most appropriate used node architectures in the SD-EON, which was also discussed in Chapter 3 (page number 29).

Figure 10.3(a) shows a typical architecture of the SD-EON, which is based on hop-retuning. In this scenario, we consider three nodes, namely node A, node B, and node C, which are connected to an SDN controller. Each node is equipped

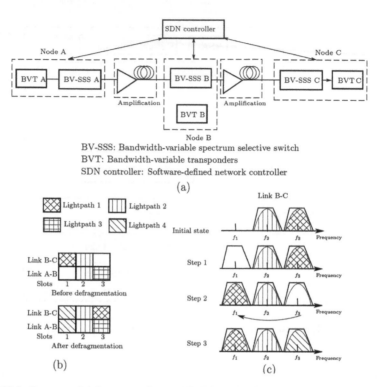

Figure 10.3: Concept of defragmentation (a) SD-EON with three nodes, (b) retuning of spectrum before and after defragmentation, (c) several defragmentation steps to accept new lightpath.

with a bandwidth-variable transponder and a BV-SSS to control traffic. When defragmentation is needed for a lightpath, the SDN controller synchronizes the intermediated nodes with corresponding equipments.

The spectrum retuning of the different lightpaths, which are lightpath 1, lightpath 2, and lightpath 3, on links A-B and B-C before and after the defragmentation is depicted in Fig. 10.3(b). Here, the spectrum retuning is based on hop-retuning, where spectrum reallocation or jump is allowed [182]. When lightpath request 4 is triggered for lightpath establishment from node A to node C, slot 1 on link A-B is available; slot 1 is unavailable on link B-C. If lightpath 1 is moved or jumped from slot 1 to slot 3 according to the hop-retuning, the spectrum resource is used efficiently. Thus, spectrum slot can be available on link B-C in order to setup new lightpath 4. Note that the spectrum retuning can also be implemented when few contiguous slots are required by a lightpath. The following technology can be implemented to achieve the above-mentioned retuning process, which does not disrupt any existing lightpath.

Several defragmentation steps with BV-SSS reconfiguration to accommodate a new lightpath, which is lightpath 4 in Fig. 10.3(b), are shown in Fig. 10.3(c). It is assumed that one slot is required for each lightpath and the central frequencies, namely f_1, f_2, and f_3, are used for slots 1, 2 and 3, respectively. Initially, before defragmentation, slot 1 on f_1 frequency is available and hence the allocated spectrum on the f_3 is retuned on f_1. Now the spectrum on f_3 is available in order to setup the new lightpath.

The SDN controller first sends control signals in order to configure the related network elements of the intermediate links, when it decides to defragment spectrum slots and retuning the lightpath from one frequency to another. Then a control information is sent by the SDN controller to intermediate nodes to conduct defragmentation.

Note that the typical architecture of the SD-EON uses a centralized controller. When defragmentation is needed for a lightpath, the SDN controller synchronizes the intermediate nodes with corresponding equipments. When natural disasters, such as earthquake or tsunami, happen, the centralized controller located in the disaster areas can be destroyed and the entire controlling system will be collapsed. The centralized control based SD-EONs is not suitable for a large network due to the scalability issue. The delay between the SDN controller and network nodes is increased when the network size increases. The reliability and scalability issues are the major limitation of the typical SD-EON architecture based on centralized controller.

To handle such issues, the idea of decentralization of the control plane [232] can be adopted for SD-EONs; flat and hierarchical SDN controlled architectures are typically adopted in the distributed SDN controlled plane. In the flat SDN controlled architecture, the entire network is horizontally partitioned into multiple areas, each of which is handled by a single controller. Whereas, the entire network control plane is vertically divided into multiple layers based on the required services in the hierarchical SDN control architecture.

10.3   Performance evaluation

This section estimates and investigates the performances of EONs in terms of blocking probability, traffic admissibility, and network's contiguous-aligned available slot ratio related to SDN functionalities. We consider the same assumptions as that of Chapter 7 (page number 85). It is assumed that δ [time unit] is required to complete the hop-retuning approach, including the synchronization of all of BV-SSSs that are involved in a lightpath. The network performance depends on the δ value. Smaller δ values indicate that SDN controllers are typically considered in a network. Note that the network performance also depends on the relationship between δ value and holding time (h). When δ is sufficiently smaller than h, δ does not have any significant effect on the network performance. The simulation results are obtained with a 95% confidence interval that is not greater than 5% of the reported average results.

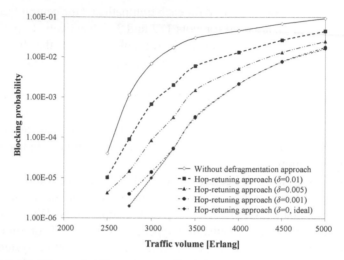

Figure 10.4: Blocking probability versus traffic volume using hop-retuning approach with different δ values.

Figure 10.4 compares blocking probabilities using the hop-retuning approach for different δ values. The blocking probability is defined as the ratio of the number of unsuccessful lightpaths to the number of considered lightpath requests in the network. We use the first fit spectrum allocation policy and the shortest path routing for each lightpath request. It is observed that the blocking probabilities increases with increase in traffic volume. For the comparison purpose, we incorporate the result of the non-defragmentation approach. We observe that when we do not consider the defragmentation approach, the network blocking probability drastically increases. We further notice that the blocking probability increases as δ value increases in the hop-retuning approach. This is because as δ value becomes large, the network requires more time to perform hop-retuning for a lightpath from one frequency to another. Typically, the larger time for hop-retuning can be suppressed by incorporating the SDN controllers. We observe that when $\delta = 0$, in an ideal condition, we achieve the best performance.

Figure 10.5 investigates how much traffic is accommodated in the network for different δ values when the acceptable blocking probability is considered as 0.01. We observe that, when δ value becomes smaller, the admissible traffic volume is increased in the network.

To evaluate the gain of adding SDN functionalities, we consider network's contiguous-aligned available slot ratio, according to [233], as one of the performance metrics. Figure 10.6 indicates that, as traffic volume increases, the contiguous-aligned available slot ratio in the network decreases. We further observe that the contiguous-aligned available slot ratio is improved when the δ value becomes smaller.

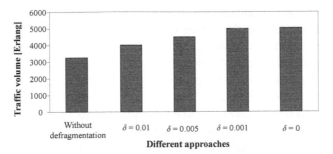

Figure 10.5: Admissible traffic for different δ values.

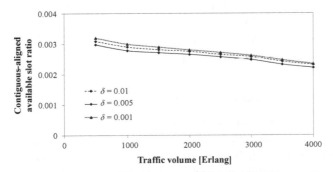

Figure 10.6: Contiguous-aligned available slot ratio using different δ values.

The above discussion summarizes that the network performance depends on δ. The incorporation of SDN architecture in the EON has a potential to make δ small. Smaller δ indicates that network resources are utilized more efficiently, which leads to suppressing the blocking probability in the network.

10.4   Related works

This section presents the related works on software-defined optical networks (SDONs), network virtualization, and different resource allocation schemes related to SD-EONs. The surveyed approaches are analyzed by clarifying their strong and weak points.

10.4.1   *Software-defined optical networks and network virtualization*

To enhance the performance of optical networks, researchers have been incorporating SDN in optical networks. In this direction, Channegowda et al. [234] presented a combined OpenFlow based control plane architecture for optical SDN networks, including an abstraction technique for enabling OpenFlow devices. Their work focuses on implementing OpenFlow protocol extensions for emerging optical transport technologies. The work in [235] summarized the optical network models and their application for the SDN management purpose.

A.S. Thyagaturu et al. [236] presented a comprehensive survey on optical networks with SDN technology. The survey begins with the introduction of software-defined optical networks (SDONs), and then focuses on different layers of the SDON. They discuss network virtualization along with the orchestration of multi-layer and multi-domain networks. The work in [237] summarized the different virtual network allocation algorithms in EONs. A survey on control plane architectures of EONs was presented in [238].

The work in [239] presents a tutorial on EONs by focusing on different research areas, which are physical layer issues, network optimization, and control plane. An in-depth survey on the current developments of SDN technology, including its three-layer architecture, is provided in [236].

10.4.2   Spectrum management

The performance of optical networks with or without SDN mainly depends on resource allocation schemes. The work in [72] provided a tutorial on RSA in EONs and its several aspects, which are modulation based quality-of-transmission, traffic grooming, fragmentation, networking cost related to RSA, energy saving, and survivability. The work in [240] presented an in-depth study on different existing RSA algorithms, and compared them in terms of resource management and computational complexity. S. Talebi et al. [73] presented a comprehensive study on spectrum allocation approaches for EONs, which analyzes and categorizes a variety of spectrum management policies, as well as fragmentation-aware RSA, distance-adaptive RSA, survivability, and traffic grooming. An exhaustive survey on spectrum management policies in EONs is available in [233], which deals with spectrum fragmentation issues. A comprehensive survey of resource allocation schemes that address different issues, namely the RSA problem in spatial mode, how to deal with crosstalk, and resource fragmentation, in space division multiplexing based optical networks was presented in [241].

A. Alyatama et al. [242] introduced an RSA algorithm using a learning approach, which is based on estimated call net gains, for EONs to enhance the network performance with respect to normalized lost revenue. The work in [243] introduced a dynamic impairment-aware spectrum allocation scheme for EONs to improve the traffic admissibility in the network.

Several studies [244–246] on the provisioning of multicast requests in WDM-based optical networks have been carried out to improve the resource utilization. Provisioning of multicast requests is still under investigation for EONs. Taking this direction, M. Moharrami et al. [247] presented a resource allocation and multicast routing scheme in EONs to suppress the call blocking. An integer linear programming (ILP) is formulated to execute both multicast routing and spectrum allocation.

The related work on SDN technologies and research issues and challenges for SD-EONs were presented in [248]. The authors in [248] emphasized the latest deployment of elastic technology for core optical networks, where generalized multiprotocol label switching (GMPLS) is executed [249]. An OpenFlow-based control plane for multilayer optical networks is presented by L. Liu et al. [250] to reduce the overall end-to-end delay.

The authors in [202] presented an optimization approach considering an SDN architecture for migrating data center services in EONs. In their scheme, the cross layer optimization and resource utilization are managed by SDN controllers according to physical layer parameters, like bandwidth demand and modulation technique. Lightpath setup and release are experimentally demonstrated through a testbed that consists with four OpenFlow-enabled EONs nodes in order to observe the call blocking and resource occupation rate.

The work in [251] described and experimentally verified online defragmentation by implying OpenFlow-assisted RSA for a single-domain SD-EON. To realize effective online defragmentation, the authors in [251] first designed the overall system by extending OpenFlow protocol and then experimented the defragmentation process that involved RSA reconfiguration. The effect of multi-domain fragmentation-aware RSA is evaluated in SD-EONs with the incorporation of OpenFlow controllers. They studied how fragmentation is managed on inter-domain links with fragmentation-aware RSA for SD-EONs.

An OpenFlow based EON architecture is demonstrated by N. Cvijetic et al. [252]. The authors in [252] extended OpenFlow 1.0 to manage optical line terminal side, which allows the instantaneous downstream communication over bandwidth-variable flex-grid channels. The introduced scheme provides low latency, high-speed, and high quality services over fiber substructures.

To maximize the admissible traffic in the network, a control-plane framework that performs online defragmentation in SD-EONs is presented by S. Ma et al. [207]. The work in [202] experimentally verified SDN over EONs for migrating data center services.

The authors in [253] demonstrated an optical channel monitoring scheme in SDN based EONs. To maintain the signal quality, the authors [253] select and adjust optical parameters dynamically.

An EON with a real-time adaptive control plane is validated by D.J. Geisler et al. [206]; an appropriate modulation format is adapted to maintain the signal quality. An SDN-based control scheme with defragmentation functionality was demonstrated to authenticate the overall possibility of extended OpenFlow messages and signal performance [208].

Note that the real deployment of the EON considering SDN is under development. We expect that fully deployed SD-EONs will be available in the near future.

Table 10.1: Summary of existing spectrum management approaches in SD-EONs.

Reference	Comments
Zhu et al. [251]	OpenFlow switches are used to validate online defragmentation
Amazonas et al. [248]	Research trends and issues for SD-EONs are presented
Liu et al. [250]	OpenFlow-based control plane is introduced to reduce delay for end-to-end lightpaths in multilayer EONs
Zhang et al. [202]	Experimentally validates SD-EONs for migration of data center services
Liu et al. [197]	EONs with OpenFlow control plane is presented
Casellas et al. [254]	How to control and manage EONs using OpenFlow controller
Cvijetic et al. [252]	Demonstration of suppressing bandwidth fragmentation using OpenFlow controller
Liu et al. [255]	Introduces a fragmentation-aware spectrum allocation approach for SD-EONs
Le et al. [256]	Introduces distributed control plane for suppressing fragmentation in dynamic SD-EONs
Zhu et al. [208]	Experimentally validates SD-EONs with defragmentation functionality

Several studies on the different spectrum management approaches in SD-EONs have been stated in the literature, which are summarized in Table 10.1.

Exercises

1. Explain SDN and network virtualization with their relationship.

2. Explain the role of OpenFlow in SDN.

3. What are the main components of SD-EONs?

4. Explain different layers of SD-EONs.

5. What are the major changes done in SD-EONs architecture over SDN architecture?

6. What are the advantages of decentralization of the control plane in SD-EONs over centralized controller based SD-EONs?

7. Why do the existing OpenFlow protocols need to be updated for SD-EONs?

Chapter 11

Mathematical Modeling for Problems in Elastic Optical Networks

An optimization problem is used to find the best solution from all feasible solutions. The best solution can be the minimum or maximum solution. An example of the former is finding the route from point A to point B that takes the minimum time. An example of the latter is determining how a production factory can maximize its profit by using limited materials. Both problems are optimization problems. An optimization problem can be solved by mathematical programming, a technique that expresses and solves problems as mathematical models. This chapter starts with a general description of optimization problems, and then presents different integer linear programming formulation for problems in elastic optical networks (EONs).

11.1 Basics of linear programming

11.1.1 Optimization problem

A person wants to travel from city A to city B. He can travel either by airplane or train. How can he travel with the minimum cost given the following conditions? The conditions are (i) condition 1: The price for a one-way ticket should not exceed 150$, (ii) condition 2: He should arrive at city B by 11:10 am, and (iii) condition 3: He should depart city A after 8:00 am. He checks the airplane and train schedules, which are mentioned in Table 11.1. The person can choose

Table 11.1: Transportation details.

Choice	Transportation	Departure time	Arrival time	Price ($)
1	Airplane	7:25 am	8:40 am	134.70
2	Airplane	9:50 am	11:05 am	136.70
3	Airplane	10:45 am	12:00 am	136.70
4	Train	7:56 am	10:36 am	138.50
5	Train	8:03 am	11:03 am	135.50
6	Train	8:20 am	10:56 am	138.50
7	Train	8:30 am	11:06 am	138.50
8	Train	8:33 am	11:30 am	135.50

one choice from eight choices by satisfying all conditions while maintaining the minimum cost. As all the prices in Table 11.1 are less than 150$, they satisfy condition 1. As for condition 2, choices 3 and 8 are not considered since they arrive after 11:10 am. For condition 3, choices 1 and 4 are not considered since their departure times are before 8:00 am, and arriving at city B at 11:03 am, and spending 135.50$.

An optimization problem consists of three components, namely decision variables, objective function, and constraints. In case of the above example, the decision variables are type of transportation, departure time, arrival time, and price. The objective function is the price. The constraints are conditions 1, 2, and 3. In the following, a mathematical model can be established that considers all three components.

Decision variables are the variables within a model that can be controlled. If there are n decision variables, they are represented as $x_1, x_2, \cdots, x_n$. Objective function is the function that we want to maximize or minimize. An objective function is written as $f(x_1, x_2, \cdots, x_n)$. If we want to maximize this function, we write as $\max_{x_1,x_2,\cdots,x_n} f(x_1, x_2, \cdots, x_n)$. If we want to minimize this function, we write as $\min_{x_1,x_2,\cdots,x_n} f(x_1, x_2, \cdots, x_n)$. Constraints are conditions or limitations of the problem.

11.1.2 Linear programming problem

A linear programming (LP) problem is an optimization problem in which the objective function and all the constraints are expressed as linear functions. Even if just one of them is not a linear function, this problem is not an LP problem. A linear function is expressed by $f(x_1, x_2, \cdots) = a_1 x_1 + a_2 x_2 + \cdots$, where $a_1, a_2, \cdots$ are constants.

Figure 11.1 shows the appearance of linear functions. In Fig. 11.1(a), there are two decision variables. The objective function and constraints are depicted by lines. In Fig. 11.1(b), there are three decision variables. The objective function and decision variables are depicted by the planes. Figure 11.2 shows an example

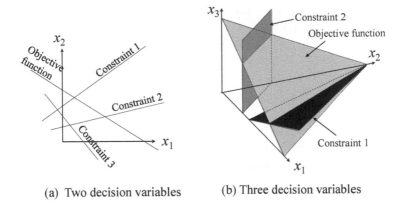

(a) Two decision variables   (b) Three decision variables

Figure 11.1: Linear programming problem.

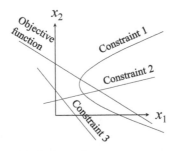

Figure 11.2: Example of non-linear programming problem.

of a non-linear programming (NLP) problem, which is not an LP problem; the objective function and constraints 2 and 3 are linear functions, but constraint 1 is not a linear function. Therefore, this problem is not an LP problem.

In general, an LP problem that minimizes an objective function is represented by the following formula.

$$\min \quad c_1 x_1 + c_2 x_2 + \cdots + c_n x_n \tag{11.1a}$$

$$\text{s.t.} \quad a_{11} x_1 + a_{12} x_2 + \cdots + a_{1n} x_n \geq b_1 \tag{11.1b}$$

$$a_{21} x_1 + a_{22} x_2 + \cdots + a_{2n} x_n \geq b_2 \tag{11.1c}$$

$$\cdots \tag{11.1d}$$

$$a_{m1} x_1 + a_{m2} x_2 + \cdots + a_{mn} x_n \geq b_m \tag{11.1e}$$

$$x_1 \geq 0 \tag{11.1f}$$

$$x_2 \geq 0 \tag{11.1g}$$

$$\cdots \tag{11.1h}$$

$$x_n \geq 0 \tag{11.1i}$$

Equations (11.1f)-(11.1i) provide the ranges of the decision variables. Equations (11.1f)-(11.1i) are not necessary for the LP problem; inclusion of (11.1f)-(11.1i) makes easy to handle the LP problem in a consistent manner. Equations (11.1a)-(11.1i) are called a canonical form of an LP problem with minimization.

An LP problem that maximizes an objective function is represented in the following.

$$\max \quad c_1 x_1 + c_2 x_2 + \cdots + c_n x_n \tag{11.2a}$$

$$\text{s.t.} \quad a_{11} x_1 + a_{12} x_2 + \cdots + a_{1n} x_n \leq b_1 \tag{11.2b}$$

$$a_{21} x_1 + a_{22} x_2 + \cdots + a_{2n} x_n \leq b_2 \tag{11.2c}$$

$$\cdots \tag{11.2d}$$

$$a_{m1} x_1 + a_{m2} x_2 + \cdots + a_{mn} x_n \leq b_m \tag{11.2e}$$

$$x_1 \geq 0 \tag{11.2f}$$

$$x_2 \geq 0 \tag{11.2g}$$

$$\cdots \tag{11.2h}$$

$$x_n \geq 0 \tag{11.2i}$$

Equations (11.2a)-(11.2i) are called a canonical form of an LP problem with maximization.

Figure 11.3 shows terminology for an LP problem with two decision variables. A boundary is a constraint that expresses the upper or lower bound of an inequality or equality. The feasible region is an area delineated by the boundaries. A corner point is an intersection of the boundaries.

The concept of LP is explained by an example. For this purpose, we consider an objective function, constraints, and two decision variables that are expressed by x_1, and x_2, which are mentioned in the following.

$$\max \quad x_1 + 5x_2 \tag{11.3a}$$

$$\text{s.t.} \quad x_1 + 10x_2 \leq 20 \tag{11.3b}$$

$$2x_1 + x_2 \leq 6 \tag{11.3c}$$

$$x_1 \geq 0 \tag{11.3d}$$

$$x_2 \geq 0 \tag{11.3e}$$

To solve an LP problem, the graphical method includes two major steps, which are (i) to determine the solution space that defines the feasible solution and (ii) to determine the optimal solution from the feasible region. Note that the set of values of the variable $x_1, x_2, x_3, \ldots x_n$, which satisfies all the constraints, and the non-negative conditions are called the feasible solution of the LP problem. Since the two decision variable x_1 and x_2 are non-negative, the first quadrant of xy-coordinate plane is only one considered. For each constraint, we draw a line.

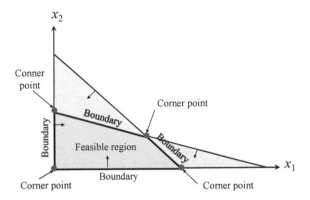

Figure 11.3: Nomenclature in LP problem.

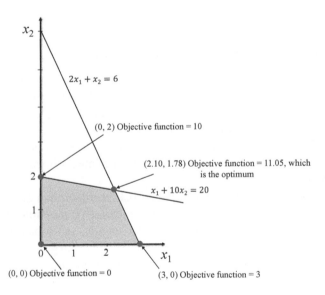

Figure 11.4: Constraints and corner points of feasible region in LP problem.

Corresponding to each constant, we obtain a shaded region. The intersection of all these shaded regions is the feasible region or feasible solution of the LP.

The above LP problem, mentioned in (11.3a)-(11.3e), is solved by the graphical method. We draw straight lines for equations $x_1 + 10x_2 = 20$ and $2x_1 + x_2 = 6$, and determine the feasible region by considering constraints, as shown in Fig. 11.4. Note that every point in the shaded region is a feasible solution of the above LP problem.

The optimal solution to an LP problem, if it exists, is found by the corner point method, which includes the following steps (i) find the feasible region of the LP problem, (ii) find the co-ordinates of each corner point of the feasible region; these co-ordinates can be obtained by solving the multiple equations provided by constraints, (iii) at each corner point, compute the value of the objective function, (iv) identify the corner point at which the value of the objective function is maximum or minimum depending on the LP problem.

Since the objective function of the above LP problem is to maximize, the optimal solution is obtained at $x_1 = 2.10$ and $x_2 = 1.78$; the optimal value for objective function is 11.05.

If the number of decision variables and the number of constraints becomes large, the complexity of obtaining all the corner points and their corresponding values of the objective function is significant. This makes the computation times so long that the solution can not be obtained in a practical time. To solve this issue, a more efficient way of finding the optimum solution for an LP problem, called the simplex method, was invented by Dantzig. The detailed information about the simplex method can be found in [257].

11.1.3 *Integer linear programming problem*

An LP problem in which decision variables take only integer values is called an Integer linear programming (ILP) problem. In LP problem, decision variables are considered to be real numbers and non-negative values. Some problems need only integer values as decision variables, such as the number of people or the number of pieces. An LP problem, in which the decision variables include both integer values and real values, is called a mixed integer linear programming (MILP) problem.

In general, it takes more time to solve an ILP problem than an LP one. ILP problems are classified into three catagories, which are (i) pure integer programming problem: all variables are required to be integer, (ii) mixed integer programming problem: some variables are restricted to be integers; the others can take any value, and (iii) binary integer programming problem: all variables are restricted to be 0 or 1.

In general, it takes more time to solve an ILP problem than an LP one. Let us consider (11.3a)-(11.3e) again, and assume that the decision variables are limited to integer values.

$$\max \quad x_1 + 5x_2 \tag{11.4a}$$
$$\text{s.t.} \quad x_1 + 10x_2 \le 20 \tag{11.4b}$$
$$2x_1 + x_2 \le 6 \tag{11.4c}$$
$$x_1 \in \{0, 1, \cdots, \} \tag{11.4d}$$
$$x_2 \in \{0, 1, \cdots, \} \tag{11.4e}$$

In an LP problem, at least one of the corner points is the optimum solution.

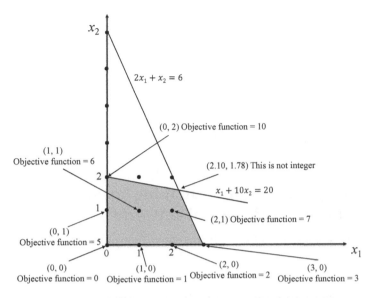

Figure 11.5: Feasible region of integer linear programming problem.

Therefore, we need check only the corner points to determine the optimum solution. However, in an ILP problem, we have to check every possible grid point in the feasible region to identify the optimum value of the objective function, as shown in Fig. 11.5. In Fig. 11.5, we need to check the value of the objective function of eight grid points, (0, 0), (1, 0), (2, 0), (3, 0), (0, 1), (1, 1), (2, 1), and (0, 2). We find that the optimum solution is (0, 2), and that the maximum value of the objective function is 10.

In the following, we consider a large-scale ILP problem with a diagram.

$$\max \quad x_1 + 5x_2 \tag{11.5a}$$
$$\text{s.t.} \quad x_1 + 10x_2 \le 2000 \tag{11.5b}$$
$$\quad 2x_1 + x_2 \le 600 \tag{11.5c}$$
$$\quad x_1 \in \{0, 1, \cdots, \} \tag{11.5d}$$
$$\quad x_2 \in \{0, 1, \cdots, \} \tag{11.5e}$$

Figure 11.6 shows that we need to find the values of the objective function by considering several grid points. The optimal solution is obtained at $x_1 = 210$ and $x_2 = 179$; the optimal value is 1105. This problem takes more time than that of the problem mentioned in (11.4a)-(11.4e) to calculate the solution since we have to consider many more grid points than those in Fig. 11.5.

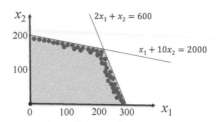

Figure 11.6: Constraints and feasible region of a large-scale ILP problem.

11.2 Integer linear programming formulation for problems in elastic optical networks

This section focusses on different ILP problems related to the spectrum resource management in EONs.

11.2.1 Creating partitioning in spectrum

This subsection discusses how to determines the number of required partitions of subcarrier slots for each fiber link [162]. Partitioning approaches for improving spectrum utilization in EONs are already discussed in Chapter 7 (section 7.1.1). Determining the number of required partitions can be expressed as a graph coloring problem [257]. The objective in partition allocation is to determine the minimum number of required partitions that accommodate all connection groups in the network with the constraint that connections assigned in the same partition must be link-disjoint. A connection group is defined as a set of connections whose routes are exactly the same. We use the term of link-disjoint connections for connections that do not share any link.

The partitioning problem is formulated as an ILP. The objective function is defined as below.

$$\min \quad n = \sum_{o \in O} y_o \tag{11.6a}$$

$$\text{s.t.} \quad \sum_{o \in O} x_v^o = 1, \quad \forall v \in V \tag{11.6b}$$

$$x_v^o + x_{v'}^o \leq y_o, \quad \forall (v, v') \in E, \forall o \in O \tag{11.6c}$$

$$y_o \geq y_{o'}, \quad \text{if } o' \geq o, \quad \forall o', o \in O \tag{11.6d}$$

$$y_o \in \{0, 1\}, \quad \forall o \in O \tag{11.6e}$$

$$x_v^o \in \{0, 1\}, \quad \forall v \in V, \quad \forall o \in O \tag{11.6f}$$

In (11.6a)-(11.6f), n, V and E are the number of required partitions, a set of vertices and a set of edges, respectively. O represents a set of colors, where $O = \{o_1, o_2, \cdots, o_{|O|}\}$. Let x_v^o and y_o be binary variables. If vertex v is assigned with color o, the value of x_v^o is 1. Otherwise, its value is 0. If o is used at least one time, the value of y_o is 1. Otherwise, the value of y_o is 0.

Equation (11.6a) expresses the objective function that minimizes the number of required partitions. It shows that the number of required partitions is equal to the total number of colors. Equation (11.6b) indicates that each vertex is assigned only one color. Equation (11.6c) ensures that two adjacent vertices must receive different colors. In other words, this constraint prevents two connection groups whose routes share the same link(s) from being assigned to the same partition. In addition, (11.6c) indicates that x_v^o must not exceed y_o for all $v \in V$. This means that if $v \in V$ such as $x_v^o = 1$ exists, y_o must be set to 1. Equation (11.6d) states that partitions are used in an ascending order of the partition index $i \in O$. Finally, the last two constraints express that x_v^o and y_o are binary variables.

11.2.2   Creating disjoint connection group

It has been observed in [161] that the allocation of disjoint connections avoids the occurrence of non-aligned available slots. Therefore, it is desirable that the disjoint connection group accommodates as large number of connections as possible. The creation of lightpath groups based on disjoint and non-disjoint routes of lightpath requests are already explained in Chapter 7 (section 7.1.1.2). The objective of creating a disjoint connection group is to maximize the total traffic demands of the disjoint connections. This optimization problem is formulated as an integer linear programming (ILP) model as below.

$$\max \quad \sum_{p \in P} g_p w_p \tag{11.7a}$$

$$\text{s.t.} \quad \sum_{j:(i,j) \in L} x_{ij}^p - \sum_{j:(j,i) \in L} x_{ji}^p = 1$$
$$\forall p = (s,d) \in P, i = s \tag{11.7b}$$

$$\sum_{j:(i,j) \in L} x_{ij}^p - \sum_{j:(j,i) \in L} x_{ji}^p = 0$$
$$\forall p = (s,d) \in P, i \neq s,d \tag{11.7c}$$

$$\sum_{p \in P} z_{ij}^p \leq 1 \quad \forall (i,j) \in L \tag{11.7d}$$

$$z_{ij}^p \leq x_{ij}^p \quad \forall (i,j) \in L, p \in P \tag{11.7e}$$

$$z_{ij}^p \leq g_p \quad \forall (i,j) \in L, p \in P \tag{11.7f}$$

$$z_{ij}^p \geq x_{ij}^p + g_p - 1 \quad \forall (i,j) \in L, p \in P \tag{11.7g}$$

$$z_{ij}^p \in \{0,1\} \quad \forall (i,j) \in L, p \in P \tag{11.7h}$$

$$g_p \in \{0,1\} \quad \forall p \in P \tag{11.7i}$$

$$x_{ij}^p \in \{0,1\} \quad \forall (i,j) \in L, p \in P \tag{11.7j}$$

In (11.7a)-(11.7j), g_p, x_{ij}^p, and z_{ij}^p are binary decision variables, and w_p is given parameter. If a path between source-destination pair p belongs to the disjoint connection group, g_p is set to 1. Otherwise its value is 0. If (i, j) is used for source-destination pair p, x_{ij}^p is 1, and otherwise 0. Lastly, if the link (i, j) is used for the source-destination pair p and is put into the disjoint connection group, z_{ij}^p is 1, and otherwise 0.

Equation (11.7a) expresses the objective function that maximizes the total traffic demands of the disjoint connection group. Equations (11.7b) and (11.7c) represent the flow constraints. Note that if the flow conservation constraints at the source and intermediate nodes are satisfied in (11.7b) and (11.7c), the flow constraints at the destination node are satisfied and do not need to be added. Equation (11.7d) ensures that the disjointness of paths in the disjoint connection group. Equations (11.7e)-(11.7g) show the relationship among variables g_p, x_{ij}^p, and z_{ij}^p. It indicates that z_{ij}^p must be equal to 1, if both g_p and x_{ij}^p are 1. Finally, the last three constraints in (11.7h)-(11.7j) are used to express the binary variables.

11.2.3   Route partitioning

It has been observed in [174] that the route partitioning (RP) optimization problem is used to minimize the end-of-line situations that can not allow a lightpath to be retuned during push-pull defragmentation process; the push-pull retuning is already discussed in Chapter 8 (section 8.1.2.2). The objective of the RP optimization problem is to minimize the total interferences among nodes sharing partitions after the cut.

A network is represented as a directed graph $G(V, E)$, where V is a set of nodes, and E is a set of links. Let P denotes the collection of all route requests $p = (s, d)$. The interference cost between routes p and q is given by w_{pq}. The outputs are the routing path x_{ij}^p and the partitions are given by d_{pq} for all routes $p = (s, d)$. d_{pq} is set to 1 if route p and q are in different partitions.

d_{pq}, y_{pq}, x_{ij}^p, k_{pq} are used as binary decision variables. d_{pq} is equal to 1 if routes p and q are in different sides of the cut, and 0 otherwise. y_{pq} is equal to 1 if routes p and q share a same link, and 0 otherwise. If link (i, j) is used for route p, then x_{ij}^p is equal to 1, and 0 otherwise; k_{pq} is equal to 1 if routes p and q share a link and are in the same side of the cut.

$$\min \sum_{p,q} (1 - d_{pq}) y_{pq} w_{pq} \qquad (11.8)$$

Since (11.8) is not a linear form, the equivalent (11.9a) is used for the ILP formulation. Variable k_{pq} is introduced with constraints (11.9g)-(11.9i) to pass from quadratic to linear formulation.

$$\min \sum_{1 \leq p < q \leq |P|} w_{pq} \times k_{pq} \qquad (11.9a)$$

s.t.
$$\sum_{j \in V:(i,j) \in E} x_{ij}^p - \sum_{j \in V:(j,i) \in E} x_{ji}^p = 1 \quad \forall p = (s,d) \in P,$$
$$i \in V, i = s \tag{11.9b}$$

$$\sum_{j \in V:(i,j) \in E} x_{ij}^p - \sum_{j \in V:(j,i) \in E} x_{ji}^p = 0 \quad \forall p = (s,d) \in P,$$
$$i \in V, i \neq s,d \tag{11.9c}$$

$$d_{pq} + d_{pk} + d_{qk} \leq 2 \quad \forall 1 \leq p < q < k \leq |P| \tag{11.9d}$$

$$d_{pq} - d_{pk} - d_{qk} \leq 0 \quad \forall 1 \leq p < q \leq |P|, k \neq p,q \tag{11.9e}$$

$$x_{ij}^p + x_{ij}^q - 1 \leq y_{pq} \quad \forall (i,j) \in E, p,q \in P \tag{11.9f}$$

$$k_{pq} \leq y_{pq} \quad \forall p,q \in P \tag{11.9g}$$

$$k_{pq} \leq 1 - d_{pq} \quad \forall p,q \in P \tag{11.9h}$$

$$k_{pq} \geq y_{pq} - d_{pq} \quad \forall p,q \in P \tag{11.9i}$$

$$x_{ij}^p \in \{0,1\} \quad \forall (i,j) \in E, p \in P \tag{11.9j}$$

$$y_{pq}, d_{pq}, k_{pq} \in \{0,1\} \quad \forall p,q \in P \tag{11.9k}$$

Equations (11.9b) and (11.9c) represent the traffic flow constraints. They ensure that all traffic leaving a source node are routed to destination node without any traffic lost. The constraint on the destination node is not added as it has been proved to be redundant when the constraints at the source in (11.9b) and the intermediate nodes in (11.9c) are stated. Equation (11.9f) defines the auxiliary graph from the routing paths. It sets y_{pq} to 1 if routes p and q share at least a link (i,j), otherwise it is forced to 0 ($x_{ij}^p = 0$ or $x_{ij}^q = 0$ on all (i,j) links). Equations (11.9d) and (11.9e), called triangle inequalities. The triangle inequalities induce facets of the cut, where every three nodes define a facet, which either does or does not intersect with the cut plane. When a facet intersects with the cut plane, two of its nodes are in one side of the cut and the last one is in the opposite side. Therefore, for the two nodes in the same side, d_{pq} is equal to zero; in (11.9d) the sum of d_{pq} is less or equal to 2. Equation (11.9e) guaranties that d_{pq} is equal to 1 for nodes in opposite sides. When a facet does not intersect with the cut plane, d_{pq} is equal to zero for all pairs, which does not contradict (11.9d) and (11.9e). When possible, facets are forced to intersect with the cut to minimize path interference. Equation (11.9f) puts y_{pq} to 1 if paths p and q share the same link. Equations (11.9g)-(11.9i) define the remaining edges after the cut. They set k_{pq} to zero if routes p and q are not connected ($y_{pq} = 0$) or are allocated to different sides of the cut ($d_{pq} = 1$).

11.2.4 Path exchanging in 1+1 protected EONs

This subsection formulates the optimization problem to minimize the spectrum fragmentation while limiting the number of network operations in 1+1 protected EONs [212]. Path exchanging operations are used during the defragmentation process that minimizes the spectrum fragmentation, which is already presented in Chapter 9. The objective function is intended to minimize the highest used index, which is the highest spectrum slot used by a lightpath on any link of the network. Minimizing it limits the spectrum fragmentation since it pushes allocated lightpaths to the lower spectrum indexes (occupied spectrum indexes do not exceed the highest used index).

With the highest used index to be minimized in the objective function, several instances of spectrum rearrangement can yield the same minimum value. In order to choose the instance that requires the least number of network operations, we affix to this objective the secondary objective which is not a primary priority. The secondary objective is to select the instance that requires the least number of combined path exchanging and backup reallocation operations from the instances that yield the minimum highest used index. Note that the secondary objective can be defined to allow the selection of relaxed solutions with instances that require low number of network operations given that they return the highest used index value close enough to the minimum value.

In this formulation, the constraints are used for the lightpath establishment, the spectrum consecutiveness and the slot capacity constraints, and the transition constraints to track and ensure the validity of the lightpath reallocation moves.

The lightpath establishment constraint ensures that all lightpaths are allocated. Each established lightpath is identified by its starting allocation index defined by the lowest spectrum slot index that it occupies. The spectrum consecutiveness is set so that lightpaths are allocated contiguous spectrum slot indexes. The spectrum slot capacity constraint prevents a spectrum slot index on a given link from being used by at more than one lightpath at the time. In other words, two lightpaths sharing a link cannot use the same spectrum slot index at the same time.

The transition constraints are introduced to track all lightpath reallocation moves during the transition period. They enforce the restrictions applying to the lightpath reallocation and the path exchanging operations, namely (i) a lightpath cannot be reallocated to slots used by another lightpath before the latter is reallocated, (ii) primary paths have to be toggled to the backup state before being reallocated, and (iii) the primary and backup lightpaths transmitting the same signal cannot be reallocated simultaneously.

It is assumed that the network's link capacity and spectrum state are given. All links ($e \in E$) are supposed to have the same capacity, which is the number of available spectrum slots $|F|$ per link, indexed from 0 to $|F| - 1$ as $F = \{0, \cdots, |F| - 1\}$. The initial spectrum state is represented by the set of lightpaths ($p \in P$) and their initial lightpath allocation indexes f_p^{init}. We refer to the set of path routed through link e as P_e. The lightpaths' signal identifier $s(p)$ are also given; a primary lightpath and its backup lightpath share the same signal identifier. The initial state of a lightpath k_p^{init} is set to 1 for primary and 0 for backup.

$n(p)$, which is a positive integer, represents the number of spectrum slots occupied by lightpath p on links.

To ensure that the returned target state can be reached using the path exchanging scheme, we use on variables the step dimension t, in addition to the paths p and indexes f, to follow the steps of the transition process. All constraints previously described must hold at each step t of the transition process. The maximum number of steps T is given as a parameter. We define $\tau = \{1, \cdots, T\}$ as the set of steps, $t = 0$ corresponds to the initial state, and $t = T$ corresponds to the returned state. We also define $\tau^{-1} = \{1, \cdots, T-1\}$.

During each step, either a swapping operation or a move operation can be performed on a given lightpath, but not both. On the other hand, multiple lightpaths can be considered in one step as long as the slot capacity and transition constraints hold. Note that the target state can be reached before running T steps. It occurs if T is larger than the number of required steps. As we limit the number of network operations with the second objective, the number of steps that it takes to reach the target state is not a concern as long as the transition constraints hold.

The used decision variables are described as follows. $U(t)$ is a positive integer defined as the highest spectrum slot occupied by any lightpath at step t. The binary variables $x_p^f(t)$ and $y_p^f(t)$ are used to identify respectively the starting allocation index f of lightpath p and a spectrum slot index f allocated to lightpath p. $x_p^f(t)$ is set to 1 if f is the lowest used index of lightpath p at step t, and 0 otherwise. $y_p^f(t)$ is set to 1 if lightpath p uses spectrum slot index f at step t, and 0 otherwise. Binary $k_p(t)$ represents the state of lightpath p to indicate the primary or backup state at step t. It is put to 1 if lightpath p is a primary lightpath and 0 if it is a backup one. The path exchanging operations are registered by $h_p(t)$, which is binary. $h_p(t)$ is set to 1 if primary lightpath p is toggled to the backup state at the end of step t, and 0 otherwise. Recall that, while a primary lightpath is toggled to the backup state during a path exchanging operation, the corresponding backup is simultaneously toggled to the primary state. Therefore, counting the number of times a primary lightpath is toggled to backup is equivalent to counting the number of path exchanging operations. $m_p(t)$ is a binary to track whether lightpath p is reallocated. It is activated to 1 if lightpath p is reallocated during the transition between step t and $t+1$, and 0 otherwise.

In the following, the ILP formulation is presented.

$$\min \quad U(T) + \varepsilon \times \sum_{t=1}^{T} \sum_{p \in P} (h_p(t) + m_p(t)) \tag{11.10a}$$

$$\text{s.t.} \quad x_p^f(0) = 1 \quad \forall p \in P, f = f_p^{\text{init}} \tag{11.10b}$$

$$k_p(0) = k_p^{\text{init}} \quad \forall p \in P \tag{11.10c}$$

$$\sum_{f \in F} x_p^f(t) = 1 \quad \forall p \in P, t \in \tau \tag{11.10d}$$

$$x_p^f(t) \leq y_p^{f'}(t) \quad \forall p \in P, t \in \tau,$$
$$f \in \{0, \cdots, |F| - n(p)\}, f' \in \{f, \cdots, f + n(p) - 1\} \quad (11.10\text{e})$$
$$x_p^f(t) = 0 \quad \forall p \in P, t \in \tau,$$
$$f \in \{|F| - n(p) + 1, \cdots, |F| - 1\} \quad (11.10\text{f})$$
$$\sum_{p \in P_e} y_p^f(t) \leq 1 \quad \forall e \in E, f \in F, t \in \tau \quad (11.10\text{g})$$
$$f \times y_p^f(t) \leq U(t) \quad \forall p \in P, f \in F, t \in \tau \quad (11.10\text{h})$$
$$|x_p^f(t) - x_p^f(t+1)| \leq m_p(t) \quad \forall$$
$$p \in P, f \in F, t \in \tau^{-1} \quad (11.10\text{i})$$
$$y_p^f(t) + \sum_{p' \in P_e : p' \neq p} y_{p'}^f(t+1) \leq 1 \quad \forall e \in E,$$
$$p \in P_e, f \in F, t \in \tau^{-1} \quad (11.10\text{j})$$
$$k_p(t) + m_p(t) \leq 1 \quad \forall p \in P, t \in \tau \quad (11.10\text{k})$$
$$k_p(t+1) + m_p(t) \leq 1 \quad \forall p \in P, t \in \tau^{-1} \quad (11.10\text{l})$$
$$k_p(t) + k_{p'}(t) = 1 \quad \forall t \in \tau$$
$$p \in P, p' \in P, p' \neq p, s(p') = s(p) \quad (11.10\text{m})$$
$$k_p(t) - k_p(t+1) \leq h_p(t) \quad \forall p \in P, t \in \tau^{-1} \quad (11.10\text{n})$$
$$x_p^f(t), y_p^f(t) \in \{0, 1\} \quad \forall p \in P, f \in F, t \in \tau \quad (11.10\text{o})$$
$$m_p(t), k_p(t), h_p(t) \in \{0, 1\} \quad \forall p \in P, t \in \tau \quad (11.10\text{p})$$

The objective to be minimized is the highest used index after the defragmentation $U(T)$. It is represented by (11.10a). The second part of (11.10a) is used to select the target spectrum that limits the number of network operations among solutions that have the same value for the first term of (11.10a). It sums the number of path exchanging operations represented with $h_p(t) = 1$ when primary paths are toggled to the backup state and the number of backup reallocation operations represented with $m_p(t) = 1$. ε is selected small enough not to impair the first term of (11.10a). Equation (11.10b) sets the initial spectrum. Equation (11.10c) initializes the lightpaths' state, primary or backup. Equation (11.10d) represents the lightpath establishment constraint, which ensures that, for each step t, lightpath p is allocated and has a unique starting allocation index. The consecutiveness constraint is expressed by (11.10e) and (11.10f). Equation (11.10e) ensures that $y_p^{f'}(t)$ for each of the $(n(p) - 1)$ spectrum slots f' following the starting allocation index f of lightpath p, for which $x_p^f(t)$ is equal to 1, is forced to 1. In other words, (11.10e) ensures that the $n(p)$ spectrum slots allocated to lightpath p are consecutive from its starting allocation index f. Equation (11.10f) ensures that there are at least $n(p)$ spectrum slots between the starting allocation index of lightpath p and the last index of the spectrum. The highest used spectrum index, $U(t)$, is returned by (11.10h). It puts $U(t)$ equal to the highest spectrum index used by any path $(\max_{p\in P} f \times y_p^f(t))$ at step t. Equations (11.10i) to (11.10m) represent the constraints to define and track the transition moves. Equation (11.10i) tracks whether a lightpath is reallocated at step t. If lightpath p is reallocated at step t, $m_p(t)$ is forced to 1. Any change between $x_p^f(t)$ and $x_p^f(t+1)$ forces $m_p(t)$ to 1 for (11.10i) to hold. In fact, for all $f \in F$, if either $x_p^f(t)$ or $x_p^f(t+1)$ is equal to 1, then the other term has to be equal to 1 for $m_p(t)$ to be equal to 0. This means, if lightpath p is moved from or to the spectrum index f, then $m_p(t)$ is 1. Otherwise, $m_p(t)$ is minimized to 0 by the second term of (11.10a). Equation (11.10j) forbids lightpaths to be reallocated to spectrum slots used by another lightpath with which it shares a link unless the latter has already been reallocated. In other words, we cannot reallocate a lightpath to spectrum slots used by another lightpath that is to be reallocated in the same step if they are not path disjoint. Equation (11.10k) forbids primary lightpaths to be reallocated and (11.10l) makes sure that reallocated lightpaths cannot be toggled right away. Thus, they ensure that primary lightpaths are toggled to backup first before being reallocated and that reallocated lightpath cannot be used in a path exchanging operation during the same step. Equation (11.10m) ensures that, while a lightpath is in the primary state, the other lightpath of the 1+1 protection is in the backup state and vice-versa. Two lightpaths transmitting the same signal cannot be reallocated at the same time for data integrity. Equation (11.10n) detects the path exchanging operations by forcing $h_p(t)$ to 1 if lightpath p is toggled from the primary state to the backup one. Otherwise, $h_p(t)$ is minimized to 0 due to the second term of the objective function. Finally, (11.10o)-(11.10p) define the binary variables.

11.2.5 Path exchanging in reroutable backup paths in 1+1 protected EONs

To enhance the performance in terms of resource utilization, Sawa et al. [230] formulated the optimization problem to minimize the highest used spectrum slot index while limiting the number of network operations in reroutable backup paths in 1+1 protected EONs considering path exchanging. The highest used spectrum slot index expresses the highest spectrum slot in all links used by every lightpath of the network. Minimizing the highest spectrum slot index can lead to pushing a lightpath with a higher index to a lower index. The network operations in the optimization problem consist of toggling the functions of primary and backup, reallocating and rerouting lightpaths to the spectrum.

Note that the optimization problem mentioned in this subsection is the extension of the optimization problem mentioned in section 11.2.4 in which the backup paths are allowable for rerouting.

For minimizing the highest spectrum slot index, the flow of the network operations from the initial state to a target state is not always one instance. To minimize the number of the network operations and get the instance with the least number, the second term of the objective function is affixed following the first term of the objective function for the highest spectrum slot index. The second term of the objective function represents the total number of toggling operations and spectrum reallocating operations.

The constraints of the ILP problem are defined as follows. The lightpath establishment constraints give an initial state including the spectrum slot indexes, the functions of primary or backup and the routes of the lightpaths. Each established lightpath is identified by the lowest spectrum slot index that the lightpath occupies. The spectrum continuity constraints require that a lightpath uses the same spectrum slot index on each link. The spectrum contiguity constraints require that a lightpath is allocated to contiguous spectrum slots on a link. The constraints of the routing prohibit a primary path and the corresponding backup path to share any common link. The constraints of the preservation of flow impose that each path has a pair of source node and destination node and a set of links make a path continuing from the source node to the destination node. The transition constraints limit network operations during the transition period. A lightpath cannot be reallocated and rerouted to the spectrum slots used by another lightpath before the latter is reallocated or rerouted to the other spectrum slots.

In the ILP problem, the initial state of spectrum and links is given. A network is expressed by directed graph $G(V, E)$, where V is the set of nodes and E is the set of directional links. (i, j) denotes a link from node $i \in V$ to node $j \in V$. F denotes the set of spectrum slot indexes from 0 to $|F| - 1$, namely $F = \{0, \cdots, |F| - 1\}$. Every link has the same capacity $|F|$. In the initial state, a spectrum slot index used by lightpath $p \in P$, where P is the set of lightpaths, is given as $f_p^{\text{init}} \in F$ and lightpath p starts from $o(p) \in V$ to $d(p) \in V$. Binary constant k_p^{init} indicates that lightpath p is a primary path or a backup path at the initial state; it is set to 1 for a primary path and 0 for a backup path. Binary constant $r_p^{\text{init}}(i, j)$ is set to 1 if the lightpath p uses link (i, j) at the initial state and 0 otherwise. A primary path and the corresponding backup path have the same signal identifier $s(p)$. The number of spectrum slots used by lightpath p is represented by $n(p)$.

In the process of the defragmentation from the initial state to the target state, the step dimension is represented by step index t and the maximum number of steps is represented by constant T. $\tau = \{0, \cdots, T\}$ denotes the set of steps; $t = 0$ expresses the initial state and $t = T$ expresses the target state. $\tau^{-1} = \{0, \cdots, T - 1\}$ is also defined. During each step, if the toggling operation is taken to a lightpath, the reallocating and rerouting operation cannot be taken to the lightpath, and vice versa. The number of network operations is reduced by the second term of the objective function and the number of steps has to be fewer than the maximum number of T steps.

The decision variables in the ILP problem are represented as follows. Non-negative integer $U(t)$ denotes the highest spectrum slot index in the all links used by every lightpath of the network at step t. $x_p^f(t)$ and $y_p^f(t)$, binary decision variables, express which spectrum slots lightpath p is allocated to at step t. If the lowest spectrum slot index used by lightpath p is equal to $f \in F$ at step t, $x_p^f(t)$ is set to 1, and otherwise 0. If the spectrum slot $f \in F$ is used by lightpath p at step t, $y_p^f(t)$ is set to 1, and otherwise 0. $r_p(t,i,j)$ is a binary decision variable for routing; if lightpath p uses link (i,j) at step t, $r_p(t,i,j)$ is set to 1, and otherwise 0. $z_p^f(t,i,j)$ is a binary decision variable, which expresses the product of $y_p^f(t)$ and $r_p(t,i,j)$ to a linear form. $w_p^f(t,i,j)$ is a binary decision variable, which expresses the product of $r_p(t-1,i,j)$ and $\sum_{p' \in P:p' \neq p} z_{p'}^f(t+1,i,j)$ to a linear form. $k_p(t)$ is a binary decision variable for the state of lightpath p; it is set to 1 if lightpath p is primary at step t, and it is set to 0 if lightpath p is backup. $h_p(t)$ is a binary decision variable for toggling operation; if lightpath p is toggled to primary or backup at the end of step t and the corresponding lightpath is toggled to the opposite state, it is set to 1, and otherwise 0. $m_p(t)$ is a binary decision variable for reallocating operation; if lightpath p is reallocated to the spectrum between step t and $t+1$, $m_p(t)$ is set to 1, and otherwise 0. $l_p(t)$ is a binary decision variable for rerouting; if lightpath p is rerouted between step t and $t+1$, $l_p(t)$ is set to 1, and otherwise 0.

In the following, the ILP formulation is presented.

$$\min \quad U(T) + \varepsilon \times \sum_{t=0}^{T-1} \sum_{p \in P} (h_p(t) + m_p(t) + l_p(t)) \tag{11.11a}$$

$$\text{s.t.} \quad x_p^f(0) = 1, \forall p \in P, f = f_p^{\text{init}} \tag{11.11b}$$

$$k_p(0) = k_p^{\text{init}}, \forall p \in P \tag{11.11c}$$

$$r_p(0,i,j) = r_p^{\text{init}}(i,j), \forall p \in P, (i,j) \in E \tag{11.11d}$$

$$\sum_{f \in F} x_p^f(t) = 1, \forall p \in P, t \in \tau \tag{11.11e}$$

$$x_p^f(t) \leq y_p^{f'}(t), \forall p \in P, t \in \tau,$$
$$f \in \{0, \cdots, |F| - n(p)\}, f' \in \{f, \cdots, f + n(p) - 1\} \tag{11.11f}$$

$$x_p^f(t) = 0, \forall p \in P, t \in \tau,$$
$$f \in \{|F| - n(p) + 1, \cdots, |F| - 1\} \tag{11.11g}$$

$$\sum_{p \in P} z_p^f(t,i,j) + \sum_{p \in P} z_p^f(t,j,i) \leq 1, \forall t \in \tau,$$
$$(i,j) \in E, f \in F \tag{11.11h}$$

$$z_p^f(t,i,j) \leq y_p^f(t), \forall p \in P, t \in \tau, (i,j) \in E, f \in F \tag{11.11i}$$

$$z_p^f(t,i,j) \leq r_p(t,i,j), \forall p \in P, t \in \tau, (i,j) \in E, f \in F \tag{11.11j}$$

$$z_p^f(t,i,j) \geq y_p^f(t) + r_p(t,i,j) - 1,$$
$$\forall p \in P, t \in \tau, (i,j) \in E, f \in F \tag{11.11k}$$

$$\sum_{j:(i,j)\in E} r_p(t,i,j) - \sum_{j:(j,i)\in E} r_p(t,j,i) = 1,$$
$$\text{if } i = o(p), \forall p \in P, t \in \tau \tag{11.11l}$$

$$\sum_{j:(i,j)\in E} r_p(t,i,j) - \sum_{j:(j,i)\in E} r_p(t,j,i) = 0,$$
$$\text{if } i \neq (o(p), d(p)) \in V, \forall p \in P, t \in \tau \tag{11.11m}$$

$$r_p(t,i,j) + r_{p'}(t,i,j) \leq 1, \text{if } p \neq p', s(p) = s(p'), \forall p \in P,$$
$$t \in \tau, (i,j) \in E \tag{11.11n}$$

$$\sum_{(i,j)\in E} r_p(t,i,j) \leq \sigma \forall p \in P, t \in \tau \setminus \{0\} \tag{11.11o}$$

$$f \times y_p^f(t) \leq U(t), \forall p \in P, f \in F, t \in \tau \tag{11.11p}$$

$$z_p^f(t,i,j) + w_p^f(t+1,i,j) \leq 1,$$
$$\forall p \in P, t \in \tau^{-1}, (i,j) \in E, f \in F \tag{11.11q}$$

$$w_p^f(t+1,i,j) \leq r_p(t,i,j), \forall p \in P, t \in \tau^{-1},$$
$$(i,j) \in E, f \in F \tag{11.11r}$$

$$w_p^f(t+1,i,j) \leq \sum_{p'\in P: p' \neq p} z_{p'}^f(t+1,i,j),$$
$$\forall p \in P, t \in \tau^{-1}, (i,j) \in E, f \in F \tag{11.11s}$$

$$w_p^f(t+1,i,j) \geq r_p(t,i,j) + \sum_{p'\in P: p' \neq p} z_{p'}^f(t+1,i,j) - 1,$$
$$\forall p \in P, t \in \tau^{-1}, (i,j) \in E, f \in F \tag{11.11t}$$

$$|x_p^f(t) - x_p^f(t+1)| \leq m_p(t), \forall p \in P, t \in \tau^{-1}, f \in F \tag{11.11u}$$

$$|r_p(t,i,j) - r_p(t+1,i,j)| \leq l_p(t), \forall p \in P,$$
$$t \in \tau^{-1}, (i,j) \in E, f \in F \tag{11.11v}$$

$$k_p(t) + m_p(t) \leq 1, \forall p \in P, t \in \tau \tag{11.11w}$$

$$k_p(t+1) + m_p(t) \leq 1, \forall p \in P, t \in \tau^{-1} \tag{11.11x}$$

$$k_p(t) + l_p(t) \leq 1, \forall p \in P, t \in \tau \tag{11.11y}$$

$$k_p(t+1) + l_p(t) \leq 1, \forall p \in P, t \in \tau^{-1} \tag{11.11z}$$

$$k_p(t) + k_{p'}(t) = 1, \forall t \in \tau, p \in P,$$
$$p' \in P, p' \neq p, s(p') = s(p) \tag{11.11aa}$$

$$k_p(t) - k_p(t+1) \leq h_p(t), \forall p \in P, t \in \tau^{-1} \tag{11.11bb}$$

$$x_p^f(t), y_p^f(t) \in \{0,1\}, \forall p \in P, f \in F, t \in \tau \tag{11.11cc}$$

$$m_p(t), k_p(t), h_p(t), l_p(t) \in \{0,1\}, \forall p \in P, t \in \tau \tag{11.11dd}$$

$$r_p(t,i,j) \in \{0,1\}, \forall p \in P, t \in \tau, (i,j) \in E \qquad (11.11\text{ee})$$

$$z_p^f(t,i,j), w_p^f(t,i,j) \in \{0,1\}, \forall p \in P,$$
$$t \in \tau, (i,j) \in E, f \in F \qquad (11.11\text{ff})$$

The first term of the objective function in (11.11a) indicates the highest spectrum slot index at the end of defragmentation. The second term indicates the total number of the network operations for the all lightpaths between step 0 and step t. The second term is needed to compare the instances that have the same highest spectrum slot index at the end of the defragmentation. The second term is multiplied by ε in order to be small enough not to affect the first objective.

The constraints (11.11b)-(11.11v) about lightpath establishment, the spectrum continuity and contiguity and the preservation of flow are presented as follows. Equation (11.11b) gives $x_p^f(0)$, the initial lightpath's spectrum state. Equation (11.11c) gives $k_p(0)$, the initial lightpath's state of primary or backup. Equation (11.11d) gives $r_p(0,i,j)$, the initial routes of lightpaths. Equation (11.11e) expresses any lightpaths have to be allocated to any spectrum at each step. Equation (11.11f) expresses any lightpath have the contiguous spectrum slot index the number of which is $n(p)$ at each step. Equation (11.11g) prohibits the lightpath's spectrum slot index to exceed the max index $|F| - 1$ at each step. Equations (11.11h)-(11.11k) prohibit multiple lightpaths to share the same spectrum slot at each step instead of the constraint of $\sum_{p \in P} y_p^f(t) r_p(t,i,j) \leq 1$ mentioned in section 11.2.4; the constraints are expressed in a linear form. Equations (11.11l) and (11.11m) express the preservation of flow of paths from the source node to the destination node at each step. Equation (11.11n) prohibits a pair of paths transmitting the same signal to use the same link in order to apply for 1+1 path protection at each step. Equation (11.11o) expresses that the maximum number of hops is limited by a positive integer, σ, except at $t = 0$. Equation (11.11p) expresses that $U(t)$ is the highest spectrum slot index in the all links used by every lightpath of the network at each step. Equations (11.11q)-(11.11t) prohibit a spectrum slot used by a lightpath at step t to be used by other lightpaths at step $t + 1$ instead of the constraint of $r_p(t,i,j) y_p^f(t) + r_p(t,i,j) \sum_{p' \in P: p' \neq p} z_{p'}^f(t+1,i,j) \leq 1$ mentioned in section 11.2.4; the constraints are expressed in a linear form. Equation (11.11u) expresses that $m_p(t)$ is set to 1 if a lightpath p is reallocated to the spectrum at step t. Equation (11.11v) expresses that $l_p(t)$ is set to 1 if a backup path is rerouted at step t.

The constraints (11.11w)-(11.11ff) express constraints of transition of operation steps. Equation (11.11w) prohibits a primary path to be reallocated at step t. Equation (11.11x) prohibits a backup path toggled to primary path at step t to be reallocated at step t. Equation (11.11y) prohibits a primary path to be rerouted at step t. Equation (11.11z) prohibits a backup path toggled to primary path at step t to be rerouted at step t. Equation (11.11aa) means that either of the two lightpaths transmitting the same signal is primary path and the other is backup path in order to apply 1+1 path protection. Equation (11.11bb) means that $h_p(t)$ is set to 1 if a lightpath p is toggled at step t. The left side of Equation (11.11bb) does not have absolute value in order to avoid double count of the number of the toggling operations. Finally Equations (11.11cc)-(11.11ff) define the decision variables as binary.

Exercises

1. Explain the difference between integer linear programming (ILP) and linear programming (LP).

2. What are the steps to solve an LP problem by using the corner point method?

3. Solve the following LP problem.

$$\max \quad 10x_1 + 15x_2 \qquad (11.12a)$$
$$\text{s.t.} \quad x_1 + 2x_2 \leq 6 \qquad (11.12b)$$
$$2x_1 + x_2 \leq 8 \qquad (11.12c)$$
$$x_1 \geq 0 \qquad (11.12d)$$
$$x_2 \geq 0 \qquad (11.12e)$$

4. Solve the following ILP problem.

$$\max \quad 10x_1 + 15x_2 \qquad (11.13a)$$
$$\text{s.t.} \quad x_1 + 2x_2 \leq 6 \qquad (11.13b)$$
$$2x_1 + x_2 \leq 8 \qquad (11.13c)$$
$$x_1 \in \{0, 1, \cdots\} \qquad (11.13d)$$
$$x_2 \in \{0, 1, \cdots\} \qquad (11.13e)$$

5. Considering Fig. 11.7, formulate an optimization problem as an LP problem to determine the shortest path between node A to node B, and solve the LP problem.

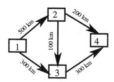

Figure 11.7: Network topology.

Chapter 12

Computational Complexity Analysis for Problems in Elastic Optical Networks

This chapter starts with a general description of computational complexity analysis of a problem, and then shows us the approach on how to prove NP-completeness of a problem. Finally, this chapter shows the proof of NP-completeness of some problems related to elastic optical networks (EONs).

12.1 Basics of computational complexity analysis

When we compare two or more algorithms, we realize that some algorithms are more efficient than others. As an efficient algorithm is preferred to solve a problem, it would be nice to have metrics for comparing algorithm efficiency. The efficiency of an algorithm is typically measured in terms of time complexity and space complexity [258, 259], which are discussed in the following.

The time complexity of an algorithm indicates the amount of time taken by the algorithm to complete its process as a function of its input size. This indicates the time of execution of the algorithm, not the time of compilation.

The space complexity of an algorithm indicates the amount of space (or memory) taken by the algorithm to run as a function of its input size. Space complexity includes both auxiliary space and space used by the input. Auxiliary space is the temporary or extra space used by the algorithm while it is being executed.

Note that, unless the space requirement of an algorithm is exponential, the time is the essential criteria to compare two or more algorithms. Thanks to the advanced memory technology, nowadays, a lot of space is available on the computer and unless an excessive amount of space is required to solve a particular algorithm, time is considered for evaluating an algorithm.

There exist several issues to estimate the time of an algorithm. Suppose, we run two algorithms, such as Algorithm A and Algorithm B, on two different machines, Machine 1 and Machine 2. If the performance of Algorithm A is better than that of Algorithm B on Machine 1 and the performance of Algorithm B is better than that of Algorithm A on Machine 2, it is difficult to say which algorithm is better in terms of time. If algorithms are measured concerning time on different machines, it becomes difficult to compare the two algorithms. Furthermore, it is observed that the performance of an algorithm in terms of time is also machine-dependent. The required times to solve the same algorithm on different machines are different.

To address the above issues, the asymptotic complexity analysis of an algorithm for both time and space is adopted. The performance of an algorithm is typically analyzed when the input size becomes large. The general steps to analyze an algorithm in terms of asymptotic notations are described in the following. (i) Find out what the input is and what the input size n of the algorithm is? (ii) determine the maximum number of operations of the algorithm in terms of n, (iii) exclude all terms, except the highest order terms, and (iv) ignore all the constant factors. The asymptotic analysis consists of the following useful notations.

12.1.1 Big O (O)

In computer science, big O notation is used to understand algorithms on how their running time or space requirements grow with increases in input size. It is typically used to provide an upper bound on the growth rate of the function. The big O notation gives us the upper bound idea, which is typically used to represent the time and space complexity of an algorithm. In the following, we formally define the big O notation.

Definition 12.1   Let $f(n)$ and $g(n)$ be two non-negative increasing functions, shown in Fig. 12.1. A function $f(n) = O(g(n))$ if there are constants, which are $c(>0)$ and $n_0 > 0$, such that $0 \leq f(n) \leq cg(n)$, for all $n \geq n_0$.

In the following, we classify algorithms from the best-to-worst performance in terms of big O notation. A logarithmic algorithm—$O(\log n)$; runtime grows logarithmically in proportion to n. A linear algorithm—$O(n)$; runtime grows di-

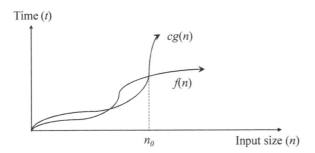

Figure 12.1: Big O notation.

rectly in proportion to n. A superlinear algorithm—$O(n\log n)$; runtime grows in proportion to n. A polynomial algorithm—$O(n^c)$; runtime grows quicker than previous all based on n. An exponential algorithm—$O(c^n)$; runtime grows even faster than a polynomial algorithm based on n. A factorial algorithm—$O(n!)$; runtime grows the fastest and becomes quickly unusable for even small values of n. Some of the examples of all those types of algorithms are mentioned as follows. Logarithmic algorithm—$O(\log n)$—binary search. Linear algorithm—$O(n)$—linear search. Superlinear algorithm—$O(n\log n)$—heap sort and merge sort. Polynomial algorithm—$O(n^c)$—strassen's matrix multiplication, bubble sort, selection sort, insertion sort, and bucket sort. Exponential algorithm—$O(c^n)$—tower of Hanoi. Factorial algorithm—$O(n!)$—determinant expansion by minors, and brute force search algorithm for the traveling salesman problem.

We provide an example for each type of algorithm considered above. Binary search—find a particular number from a list of given integer numbers that are sorted. Linear search—find a particular number from a list of given integer numbers that are unsorted. Merge sort—divide an unsorted list into sublists, each of which contains one element (a list of one element is considered sorted), and then repeatedly merge a pair of sublists to provide a new sorted sublist until there is only one sublist remaining. Strassen's matrix multiplication—given two square matrices A and B of size $n \times n$ each, find their multiplication matrix. Tower of Hanoi—given a game board with three pegs and a set of disks of different diameters that are stacked from the smallest to the largest on the leftmost peg, move all the disks to the rightmost peg following the two rules, which are (i) only one disk is moved at a time and (ii) a larger diameter disk is not placed on a smaller disk. Brute force search algorithm for the traveling salesman problem—given a list of cities and the distances between each pair of cities, what is the shortest possible route (optimum solution) that visits each city and returns to the origin city.

Theorem 12.1
$100n + 10000$ *is in* $O(n)$.

Proof 12.1   We select $n_0 = 1$ and $c = 10100, \forall n \geq n_0$.

$$100n + 10000 \leq 100n + 10000n \text{ as } n \geq 1$$
$$= 10100n$$
$$= cn$$

Thus, by definition of big O, $100n + 10000$ is in $O(n)$.   ■

12.1.2   Big Omega (Ω)

Big Ω notation provides an asymptotic lower bound. Note that the best case performance of an algorithm is typically not useful. Therefore, big Ω notation is the least used notation among all three notations. In the following, we formally define the big Ω notation.

Definition 12.2   Let $f(n)$ and $g(n)$ be two non-negative increasing functions, shown in Fig. 12.2. A function $f(n) = \Omega(g(n))$ if there are constants, which are $c(> 0)$ and $n_0 > 0$, such that $0 \leq cg(n) \leq f(n)$, for all $n \geq n_0$.

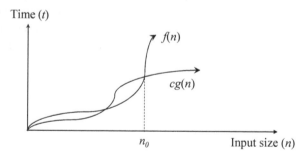

Figure 12.2: Big Ω notation.

Theorem 12.2
$2n^3 - 7n + 1$ *is in* $\Omega(n^3)$.

Proof 12.2   We select $n_0 = 3$ and $c = 1, \forall n \geq n_0$.

$$2n^3 - 7n + 1 = n^3 + (n^3 - 7n) + 1 \text{ as } n \geq 3$$
$$\geq n^3$$
$$= cn^3$$

Thus, by definition of big Ω, $2n^3 - 7n + 1$ is in $\Omega(n^3)$.   ■

12.1.3 Big Theta (Θ)

The big Θ notation bounds a function from above and below, so it defines exact asymptotic behavior. In other words, to prove big Θ notation, just prove both big O and big Ω, separately. The big Θ notation gives us the average case idea. In the following, we formally define the big Θ notation.

Definition 12.3   Let $f(n)$ and $g(n)$ be two non-negative increasing functions, shown in Fig. 12.3. A function $f(n) = \Theta(g(n))$ if there are constants, which are $c_1(> 0), c_2(> 0)$, and $n_0 > 0$, such that $0 \leq c_1 g(n) \leq f(n) \leq c_2 g(n)$, for all $n \geq n_0$.

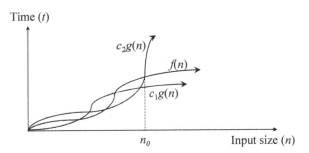

Figure 12.3: Big Θ notation.

Theorem 12.3
$2n^3 - 7n + 1$ *is in* $\Theta(n^3)$.

Proof 12.3   Theorem 12.2 proved that $2n^3 - 7n + 1$ is in $\Omega(n^3)$, when $n_0 = 3$ and $c_1 = 1$. To check the upper bound, we consider the following.
We select $n_0 = 1$ and $c_2 = 3, \forall n \geq n_0$.

$$2n^3 - 7n + 1 \leq 2n^3 + n^3 \text{ as } n \geq 1$$
$$= 3n^3$$
$$= c_2 n^3$$

By definition of big O, $2n^3 - 7n + 1$ is in $O(n^3)$.
Therefore, by definition of big Θ, $2n^3 - 7n + 1$ is in $\Theta(n^3)$.   ■

12.2 NP-completeness

The nondeterministic polynomial time (NP)-complete problems [258–260] are in NP class [260]. A problem in NP class is a decision problem, where it is verified whether an instance of the problem is a feasible solution in polynomial time. A problem X in NP class is said to be an NP-complete problem, if all other problems in NP class can be transformed (or reduced) into X within polynomial time. Note that it is unknown whether a problem in NP class can be solved within polynomial time—this is called the P versus NP problem, which is one of the million dollar problems [261].

The formal definition of NP-completeness [259,260] is as follows. A problem C is said to be an NP-complete problem, if (i) $C \in$ NP and (ii) for all $Y \in$ NP, Y is a polynomial time reducible to C. In other words, every problem in NP can be polynomially reduced to C.

12.2.1 Approach to prove NP-completeness

This subsection describes a general strategy for proving a new problem as an NP-complete problem according to [259]. Suppose that C is a given new problem. In the following, we describe the steps on how to prove problem C is an NP-complete problem.

1. Prove that $C \in$ NP.

2. Select a problem D that is a well-known NP-complete problem.

3. Consider an arbitrary instance S_D of problem D, and show how to construct, in polynomial time, an instance S_C of problem C that satisfies the following properties.

    (a) If S_D is a "yes" instance of D, then S_C is a "yes" instance of C.

    (b) If S_C is a "yes" instance of C, then S_D is a "yes" instance of D.

In other words, this establishes that S_D and S_C have the same answer.

12.3 Proof of NP-completeness

This section presents the proof of NP-completeness of some well known problems. First, we start with the graph coloring problem and then move to some problems related to optical networks.

12.3.1 3-coloring

In the 3-coloring problem, assign each node of a graph with one of three different colors under the condition that a pair of adjacent nodes connected by an edge does not use the same color. The 3-coloring problem is explained in Fig. 12.4. Nodes B and C are adjacent to node A. Therefore, each color of nodes B and C is different from that of node A. As nodes B and C are not adjacent each other, the colors of nodes B and C are the same.

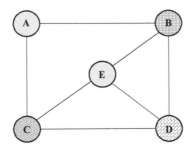

Figure 12.4: Example of 3-coloring.

Definition 12.4 Given an undirected graph $G(V,E)$ and three different colors. Does graph $G(V,E)$ 3-colorable?

Theorem 12.4
The 3-coloring decision problem (3CD) is NP-complete [259].

Proof 12.4 The 3CD problem is in NP, as it can be verified in polynomial time that at most three colors are used and no pair of nodes joined by an edge receives the same color for a given instance of 3CD problem. The time complexity to verify it is $O(|V|^2)$.

To show that the three satisfiability (3-SAT) problem, a well known NP-complete problem [262], is polynomial time reducible to 3CD. The 3-SAT problem is defined as: Given a set of n boolean variables, and k clauses of three elements each, is there a truth assignment that satisfies all clauses? Note that the 3-SAT problem is divided to clauses such that every clause contains of three literals. For example, $(x_1 \lor x_2 \lor \bar{x}_3) \land (\bar{x}_1 \lor \bar{x}_3 \lor x_4)$ is a boolean expression in 3-SAT with two clauses, each of which contains of three literals. The 3-SAT problem is: is there such values of x_1, x_2, x_3, and x_4 so that the given boolean expression is True?

An instance of 3CD is constructed from any instance of 3-SAT in polynomial time in the following. Let ϕ be a 3-SAT instance and $C_1, C_2, \cdots, C_k$ be the clauses of ϕ defined over the variables $\{x_1, x_2, \cdots, x_n\}$. The graph $G(V,E)$ is constructed for the following purposes: (i) to establish the truth assignment for $x_1, x_2, \cdots, x_n$ via the colors of the vertices of G and (ii) to capture the satisfiability of every clause C_i in ϕ.

To achieve the above two goals, a triangle with three vertices $\{T, F, B\}$ is created in G; T, F, and B represent True, False and Base, respectively. Since this triangle is a part of G, three colors are required to color G. Next, two vertices v_i and v_i' are added for every literal x_i, which creates a triangle among B, v_i, and v_i' for every (v_i, v_i') pair, as shown in Fig. 12.5(a). v_i and v_i' correspond to x_i and x_i', respectively. Therefore, the construction is not completed.

Note that the construction in Fig. 12.5(a) captures the truth assignment of the literals. Since G is 3-colorable, either v_i or v_i' gets the color T. Now, it is required to

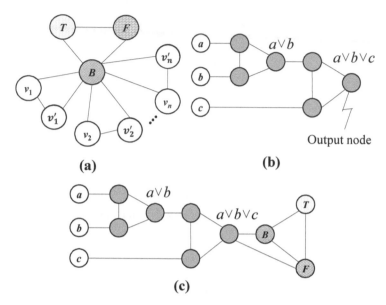

Figure 12.5: (a) Beginning of reduction for 3-coloring, (b) clause satisfiability gadget (OR-gadget), and (c) attaching C_i to OR-gadget.

add the Clause Satisfiability Gadget (OR-gadget) (see Fig. 12.5(b)) to G in order to capture the satisfiability of the clauses of ϕ.

For a clause $C_i = (a \vee b \vee c)$, the OR-gadget is expressed of its literals using T, F, and B. The gadget graph acts as a circuit whose output is mentioned at the node labeled as $a \vee b \vee c$. This node to be colored T if C_i is satisfied and F otherwise. The gadget graph is a two step construction: (i) the node labelled $a \vee b$ captures the output of $(a \vee b)$ and (ii) the same operation is repeated for $((a \vee b) \vee c)$. In the following, the gadget satisfies the following properties based on some assignments to a, b, and c.

(i) If a, b, and c are colored F in a 3-coloring, the output node of the OR-gadget to be colored F, which indicates the unsatisfiability of clause $C_i = (a \vee b \vee c)$.

(ii) If at least one of a, b, and c is colored T, there exists a valid 3-coloring of the OR-gadget, the output node of the OR-gadget to be colored T, which indicates the satisfiability of clause $C_i = (a \vee b \vee c)$.

The OR-gadget of every C_i is added to ϕ in order to connect to the output node of every gadget to the Base vertex and to the False vertex of the initial triangle, as shown in Fig. 12.5(c). In the following, it is proved that the initial 3-SAT instance ϕ is satisfiable if and only the graph G as constructed above is 3-colorable.

Suppose ϕ is satisfiable and let $(x_1^*, x_2^*, \cdots, x_n^*)$ be the satisfying assignment. If x_1^* is assigned True, v_i is colored with T and v_i' is colored with F; this is a valid coloring. Since ϕ is satisfiable, every clause $C_i = (a \vee b \vee c)$ must be satisfiable; at least one of a, b and c is set to True. The second property of the OR-gadget indicates that the gadget corresponding to C_i can be 3-colored, and hence the output node is colored as T. As the output node is adjacent to the False and Base vertices of the initial triangle, this is a proper 3-coloring.

Conversely, suppose G is 3-colorable. This constructs an assignment of the literals of ϕ by setting x_i to True if v_i is colored T and False otherwise. Now suppose this assignment is not a satisfying assignment to ϕ, which means there exists at least one clause $C_i = (a \lor b \lor c)$ is not satisfiable. This indicates that a, b, and c are set to False. If this is the case, then the output node of corresponding OR-gadget of C_i must be colored F, according to property (i). However, this output node is adjacent to the False vertex colored F, which contradicts the 3-colorability of G. ■

Note that, when $m > 3$, the 3-coloring problem can be polynomially reduced to the m-coloring problem. An instance of 3-coloring, represented by a graph G, is added to $m - 3$ new nodes. These new nodes are joined to each other and to every node in G. The resulting graph is m-colorable if and only if the original graph G is 3-colorable. Therefore, the m-coloring problem for any $m > 3$ is also NP-complete.

12.3.2 Static lightpath establishment

The establishment of lightpath using static traffic assumption is known as static lightpath establishment (SLE) problem [263]. Here, an attempt is made to minimize the number of wavelengths required to setup a given set of lightpaths under the constraint of wavelength continuity constraint, which is already discussed in Chapter 1 (section 1.2.3.2).

Definition 12.5 Given a network $G(V, E)$, a predefined set of lightpaths L, and a set of wavelengths W, is it possible to establish all lightpaths in L when $|W| \geq 3$?

Theorem 12.5
The SLE decision (SLED) problem is NP-complete [264].

Proof 12.5 The SLED problem is in NP, as it can be verified in polynomial time that a given lightpath instance of SLED is established using one of $|W|$ wavelengths. For each lightpath, a used wavelength is marked for each edge in $O(|W||E|)$. At the end of this process, it can be verified if any wavelength in each edge used by more than one lightpath in $O(|L||W||E|)$.

In the following, it is shown that the m-coloring problem, a well known NP-complete problem [262], is polynomial time reducible to SLED. The m-coloring problem is defined as: Given an undirected graph and a number m, is it possible to color the graph with at most m (where $m \geq 3$) colors such that no adjacent vertices connected by an edge in the graph receives the same color?

An instance of SLE is constructed with $|W| = m$, from any instance of the m-coloring problem in polynomial time in the following. Given a graph $G_c(V_c, E_c)$, we translate the m-coloring problem into SLED in a network $G'(V', E')$ as follows:

(i) Create node v_i^0 for every node $i \in V_c$.

(ii) For every edge $e = (i, j) \in E_c$: create four new nodes $x_{ij}, y_{ij}, v_i^r, v_j^l$ and directed edges $v_i^{r-1} \rightarrow x_{ij}, v_j^{l-1} \rightarrow x_{ij}, x_{ij} \rightarrow y_{ij}, y_{ij} \rightarrow v_i^r, y_{ij} \rightarrow v_j^l$. r and l, which are initialized to 0 in step (i), are incremented by one for v_i^r and v_j^l, respectively, when edge $e = (i, j)$ is considered.

The complexity of the above algorithm is $O(|E_c|)$. According to the above algorithm, the translation of m-coloring into a lightpath establishment using a three node graph is explained in Fig. 12.6. The three node graph for which the m-coloring problem is to be solved is shown Fig. 12.6(a). Figure 12.6(b) explains the translation of the m-coloring problem mentioned in Fig. 12.6(a) to network $G'(V', E')$. The set of lightpaths L is defined by $|V_c|$ where lightpath i uses of all links marked with i. Note that, in $G'(V', E')$, v_1^1 and v_1^2 are the first and second replications of node 1 of graph $G_C(V_c, E_c)$, v_2^1 and v_2^2 are the first and second replications of node 2 of graph $G_c(V_c, E_c)$. v_3^1, and v_3^2 are the first and second replications of node 3 of graph $G_c(V_c, E_c)$.

To conclude that the SLED instance is a feasible instance, it can be verified that the routes of all lightpaths in L exist in the network $G'(V', E')$. This is true, since

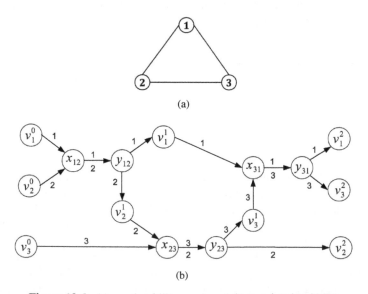

(a)

(b)

Figure 12.6: (a) n-colorability graph and (b) translated to SLED.

the transformation algorithm defines a unique route for each lightpath. Thus, if the
m-coloring problem instance is feasible, then the SLED instance is feasible.

Finally, to complete this proof, it is observed that if an SLED instance in
$G'(V', E')$ is feasible, then the corresponding m-coloring problem instance in
$G_c(V_c, E_c)$ is feasible. ■

12.3.3 Routing and spectrum allocation

The routing and spectrum allocation (RSA) problem consists of two phases,
which are routing and spectrum allocation; RSA is already discussed in Chapter 4. If the routing is already known or predetermined, the RSA problem turns
out to be the static spectrum allocation (SSA) problem. In the following, we show
the NP-completeness of the SRA problem, and therefore the SSA problem is also
NP-complete.

Definition 12.6   Given a network $G(V, E)$, a predefined set of lightpaths L, and a
set of spectrum slots in each link S, is it possible to establish all lightpaths in L?

Theorem 12.6
The SSA decision (SSAD) problem is NP-complete .

Proof 12.6
   The SSAD problem is shown in NP in the same way of SLED. Consider
that each lightpath request requires one slot. In this case, SSAD is equivalent to SLED. SLED is a subset of SSAD. Thus, the SSAD problem is NP-
complete. ■

12.3.4 Route partitioning

It has been observed in [174] that the route partitioning (RP) problem is used to
minimize the end-of-line situations that can not allow a lightpath to be retuned
during the push-pull defragmentation process; the push-pull retuning is already
discussed in Chapter 8 (section 8.1.2.2). The RP decision problem (RPD) is defined as follows:

Definition 12.7   Given a network $G(V, E)$, a set of route requests P, and a real
number h, is it possible to define all routes in P through $G(V, E)$ and partition them
with at most h total interference?

Theorem 12.7
RPD is NP-complete [174].

Proof 12.7 RPD is in NP, as we can verify in polynomial time that a given routing and partitioning instance of the RPD has at most h total interference. The time complexity to verify it is $O(|P|^2)$.

We show that the maximum cut decision problem (max-cut), a well known NP-complete problem [262], is polynomial time reducible to RPD. Max-cut is defined as: is there a bipartition of the nodes in a given graph $G_c(V_c, E_c)$ such that the total capacity of edges with two endpoints in different sets is at least k, which is a given number?

First, we construct an instance of RPD from any instance of max-cut in polynomial time. An instance of max-cut is expressed by $G_c(V_c, E_c)$, k, w_{pq}, y_{pq} and d_{pq}, where $(p, q) \in E_c$. w_{pq} is the link capacity. y_{pq} is equal to 1 if nodes p and q adjacent, and 0 otherwise. d_{pq} is set to 1 if nodes p and q are on different sides of the cut, and 0 otherwise.

We apply the idea of the transformation in [264], which translates any instance of the graph coloring problem to an instance of the static lightpath establishment problem. In our case, we seek the cut capacity, which is the total capacity of edges with two endpoints in different sets, instead of the number of colors. Using the same polynomial time algorithm as in [264], we translate any graph $G_c(V_c, E_c)$, for which we seek the cut capacity, into a network $G'(V', E')$ in which we apply RPD as follows (See an example in Fig. 12.7):

(i) Create node v_i^0 for every node $i \in V_c$.

(ii) For every edge $e = (i, j) \in E_c$: create four new nodes $x_{ij}, y_{ij}, v_i^r, v_j^l$ and directed edges $v_i^{r-1} \to x_{ij}, v_j^{l-1} \to x_{ij}, x_{ij} \to y_{ij}, y_{ij} \to v_i^r, y_{ij} \to v_j^l$. r and l, which are initialized to 0 in step (i), are incremented by one for v_i^r and v_j^l, respectively, when edge $e = (i, j)$ is considered.

The complexity of this algorithm is $O(|E_c|)$.

Every node on $G_c(V_c, E_c)$ is translated, for RPD, into a route $p = (v_i^0, v_i^n)$, where n is the number of nodes adjacent to node p on $G_c(V_c, E_c)$. If two nodes p and q on $G_c(V_c, E_c)$ are adjacent, the corresponding routes share a link in the translated network $G'(V', E')$. If the nodes are separated on different sides of the cut on $G_c(V_c, E_c)$, the corresponding routes p and q are allocated on different partitions in $G'(V', E')$. Therefore, $y_{pq} = 1$ if routes p and q share a link, and $d_{pq} = 1$ if routes p and q are allocated on different partitions. The interference between routes p and q is set to w_{pq} if routes p and q share a link, and 0 otherwise. To complete the instance of RPD, we define h' as

$$h' = \sum_{(p,q) \in E_c} w_{pq} \quad - \quad k. \tag{12.1}$$

In the following we show that with this transformation algorithm, if a max-cut instance is feasible, then the corresponding RPD instance is feasible.

Suppose a max-cut instance is a feasible instance. It means

$$\sum_{(p,q) \in E_c} d_{pq} w_{pq} \geq k. \tag{12.2}$$

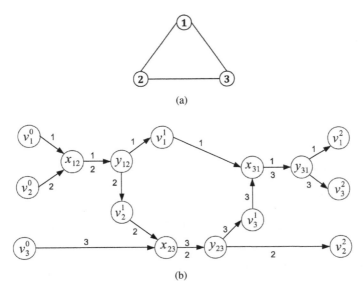

Figure 12.7: (a) Max-cut graph and (b) translated to route partitioning network.

If we subtract the terms of this equation from $\sum_{(p,q)\in E_c} w_{pq}$, then we have

$$\sum_{(p,q)\in E_c} (1-d_{pq})w_{pq} \leq \sum_{(p,q)\in E_c} w_{pq} - k$$
$$= h'. \qquad (12.3)$$

Since y_{pq} is equal to 1 if $(p,q) \in E_c$ and 0 otherwise, $y_{pq}w_{pq}$ is equal to w_{pq} if $(p,q) \in E_c$ and 0 otherwise. Therefore, we can replace w_{pq}, where $(p,q) \in E_c$, by $y_{pq}w_{pq}$, where $p,q \in P$. This gives us

$$\sum_{p,q\in P} (1-d_{pq})y_{pq}w_{pq} \leq h'. \qquad (12.4)$$

The left hand side of Eq. (12.4) is the total interference after partitioning in $G'(V',E')$, translated from $G_c(V_c,E_c)$. Therefore, Eq. (12.4) establishes that if the max-cut instance is a feasible instance, then the total interference of the RPD instance is at most h'. To conclude that the RPD instance is a feasible instance, we verify that the routing of all routes in P exist in the network $G'(V',E')$. This is true, since the transformation algorithm defines a unique routing for each route request. Thus, if the max-cut instance is feasible, then the RPD instance is feasible.

Finally, to complete this proof, we show that if a RPD instance in $G'(V',E')$ is feasible, then the corresponding max-cut instance in $G_c(V_c,E_c)$ is feasible. Suppose that an instance of RPD is feasible in $G'(V',E')$, then Eq. (12.4) is verified. By taking the opposite direction of the previous approach to verify that Eq. (12.2) leads to Eq. (12.4), we show that Eq. (12.4) leads to Eq. (12.2). With Eq. (12.4) verified, the max-cut instance is feasible in $G_c(V_c, E_c)$. Thus, if the RPD instance in $G'(V', E')$ is feasible, then the max-cut instance in $G_c(V_c, E_c)$ is feasible. ∎

Exercises

1. Prove that $5n^2 - 3n + 20$ is in $O(n^2)$.

2. Prove that $n^3 - 7n + 1$ is in $\Omega(n^3)$

3. Prove that $3n + 2$ is in $\Theta(n)$.

4. Prove that $7n + 8$ is in $\Theta(n)$.

5. What are the steps involved to analyze an algorithm in terms of asymptotic notations?

6. Give an example of a problem that is NP but not NP-complete.

7. Describe how to determine if a problem is in NP class?

8. Describe how to prove a problem is NP-complete?

9. Prove that the static spectrum reallocation for limited network operations (SSR-LNO) problem in 1+1 protected EONs problem is NP-complete.

10. Give an example of the types of algorithms whose computational time complexities, in terms of Big O notation, are logarithmic, linear, superlinear, polynomial, exponential, and factorial.

Chapter 13

Research Issues and Challenges in Elastic Optical Networks

The elastic optical network (EON) is a promising concept but its implementation remains some way off. There are several issues and challenges, which need further research to resolve [239, 265]. This chapter addresses research issues and challenges faced by optical network researchers and shows some directions for further research.

Figure 13.1 summarizes the different areas demanding further work. In the following, we identify some interesting research opportunities for the EON.

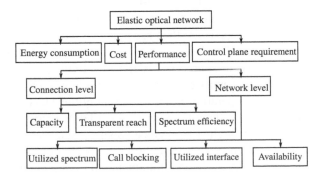

Figure 13.1: Different research areas in elastic optical networks.

13.1   Hardware development

Innovative and sophisticated devices and components must be developed in order to achieve high capacity spectrum efficient elastic optical networks. Novel optical switching and filtering elements need to be developed in order to provide efficient client protocol data unit mapping procedures that extract the incoming client signal via client-specific physical coding sublayers and media access controller layers, high resolution and steep filtering performance, optimum modulation format for bandwidth variability and higher nonlinear impairment tolerance, etc.

One of the important challenges faced by the research community is to develop a new sliceable bandwidth-variable transponder, which supports sliceability, multiple bit rates, multiple modulation formats, and code rate adaptability. However, 100 Gb/s orthogonal frequency-division multiplexing (OFDM) transponders can adapt to lower bit rates. Therefore, fractional-bandwidth services may be provided with the use of the same device. This kind of technology and the use of optical integrated circuits may offer compact and cost-effective implementation. Similarly, the need for multiple temperature-stabilized, frequency controlled lasers can be improved by phase-locked carrier generation from a single laser source.

Recently, space division multiplexing (SDM) technology [266–268] has been incorporated into elastic optical networks in order to develop high-capacity, next-generation, and few-mode/multicore fiber infrastructures. The realization of this type of infrastructure should be enabled by the development of novel multi-dimensional spectral switching nodes, which can be fabricated by extending the designs of existing flexible spectrum selective switch (SSS) nodes, through the addition of advanced mode/core adapting techniques.

To achieve a long transmission reach, optical signals must be amplified at periodic regeneration points along the fiber span to compensate the power loss experienced in multi-core fiber. For regeneration, one technique is to demultiplex the SDM signals into multiple single-core fibers and then amplify the signals in each fiber using conventional single-core erbium-doped fiber amplifiers (EDFAs). The amplified signals are then recombined and injected back into the multi-core fiber span, which increases the system delay. Therefore, to develop a single-core power transient-suppressed EDFA (TS-EDFAs) is one of the challenging research goals that must be accomplished; a key issue is to reduce the time taken to adjust the operation point of the amplifier since a newly added signal may suddenly change the total power at the EDFA input.

The development of high-performance-sophisticated bandwidth variable transponders is essential in order to synchronize transmitter and receiver ends during hitless defragmentation. In addition, bandwidth-variable optical switches and filtering components need to be developed to execute efficient protocols and support finer granularity system. Without high-performance-sophisticated devices and components, it is impossible to perform hitless defragmentation for finely-granular systems and suppress fragmentation; regardless of developments hop retuning is not possible in finely-granular systems, such as 2.5 GHz systems, due to the lack of adequate filtering components. Therefore, development of hardware and optical devices is one of essential research topics, and must be emphasized.

13.2   Network control and management

Traditional optical networks use a set of protocols (such as—generalized multiprotocol label switching (GMPLS), open shortest path first (OSPF), and resource reservation protocol - traffic engineering (RSVP)) for network control and management. Although these control protocols are well designed and standardized for traditional optical networks, evaluations that encompass the elastic technology are still at an early stage. The control plane of the elastic optical network must support many unique properties, such as—(i) optical channels are allowed to be flexible in size or width, (ii) optical channels can support various modulation formats, (iii) sub and super wavelength concept, (iv) support of multi-path routing of the composing waveband members of a split spectrum super-channel, (v) fast restoration upon failure, with or without resource reservation, etc. Therefore, new control protocols need to be developed or the existing protocols should be extended in order to support these unique properties of the elastic optical network.

During the last decade, the GMPLS protocol has been broadly standardized in different aspects including packet switching, label-switched paths (LSPs), and time-division multiplexing. However, its applicability to the elastic optical network has not been completely described yet. For this, GMPLS needs to address frequency slots instead of wavelengths. The control plane protocols are required to maintain coherent global information representing up to 320 or 640 slots, for 12.5 GHz or 6.25 GHz slot granularity, respectively [269]. The work in [270] indicated in their research work that spectrum granularity may become even finer reaching 3 GHz, resulting in 1280 possible slots. Some research works [269–271] have addressed management of the control plane. Extensive research is needed to adequately manage the control plane of the EON.

From the operation and control viewpoint, extending software-defined networking (SDN) toward transport networks, while retaining flexibility, is a challenging issue that needs to be addressed properly. In this direction, the Open Networking Foundation group [272] are addressing SDN and OpenFlow-based control capabilities for optical transport networks. In their research work, many activities such as—(i) identifying use cases, and (ii) defining reference architecture in order to control optical transport networks by incorporating the OpenFlow standards, have been performed to develop OpenFlow protocol extensions.

Based on their model, the extended OpenFlow protocol is responsible for interfacing with network elements. The control virtual network interface is developed in order to provide the required bridge between the data center controllers. Future extensions and additional standardization activities are needed in order to realize the SDN controllers that can manage TDM circuit and wavelength-based architecture (such as—generic packet/TDM/fiber switching, bandwidth aggregation and segmentation). Finally, an efficient technology needs to be developed in order to support a combination of centralized and distributed control of a multi-layer network.

A control plane that includes synchronization during the defragmentation process, is required to enhance the performance of a network. The control plane is responsible for exchanging the bandwidth profile information among nodes over a lightpath. It is also responsible for establishing, releasing, and allocating bandwidth of lightpaths. It can be centralized or decentralized. As the control plane of EONs must support many unique properties, new control protocols need to be developed or the existing protocols need to be modified in order to support the unique properties of EONs.

13.3   Energy consumption

The increase in the traffic in carried by optical networks will increase the energy consumption. The elastic optical network has the ability to significantly reduce energy consumption. In combination with sliceable bandwidth-variable transponders (SBVTs), the elastic optical network presents some new features as regards optical traffic grooming and optical layer bypass [58], which can help to reduce the energy consumption. However, we still are unable to groom traffic optically in a very early stage, and this omission must be tackled.

The elastic optical network offers a lower blocking probability compared to traditional optical networks and so can accept higher volumes of traffic. This clearly is a significant advantage in terms of energy efficiency, as the deployment of additional network elements would not only increase cost, but also increase the overall energy consumption. Several energy saving schemes are anticipated, setting some the network elements into sleep mode when the traffic is below a certain threshold. Another interesting topic for future researchers is to analyze the energy efficiency of new protection and restoration schemes for the elastic optical network.

13.4   Physical layer impairments

As optical connections may span over many long links, physical layer impairments (PLIs), such as—dispersion, interference, noise, and nonlinear effects accumulate and degrade the signal quality, which affects the quality-of-transmission (QoT). Accounting for PLIs is a challenging issue for network designers, especially if we consider exact models and the interdependencies. Many studies [87,273–275] on PLIs have been carried out for wavelength division multiplexing (WDM) based optical networks. PLIs have a distinct impact on both WDM-based optical networks and elastic optical networks. With the introduction of coherent detection and digital signal processing, impairments that are related to dispersion can be substantially reduced or fully compensated. However, high levels of flexibility make the minimization of these effects more complicated from an algorithmic perspective, which needs further research.

13.5 Spectrum management

One of the most important problems in elastic optical networks, both for planning and operation, is allocating network resources in a dynamic environment. The problem of establishing connections in fixed-grid WDM-based networks and the allocation of network resources is well addressed in the literature [6,7]. However, connection establishment in the elastic optical network is more complicated for several reasons. First, in contrast to WDM networks where each connection is assigned a single wavelength, in elastic optical networks, spectrum slots can be allocated in a flexible manner. Apart from the difference in spectrum resource allocation, the choice of the transmission parameters of the tunable transponders present in flexible networks directly or indirectly impacts the resource allocation decision and makes the problem even more complicated.

Some proposals [100] found in the literature, partition the entire spectrum in an advance in order to handle spectrum resources efficiently. Most of these schemes assume the traffic demand in advance. However, considering the fact that the traffic profile will change over time, connections with larger size slots demand will have to be accommodated over the same partitions. In this case, it is obvious that there is no other way than dropping these connections. As a result, the blocking performance will decrease significantly. Even worse, larger slot connections are the ones that will be dropped most often. Therefore, it is very important to take account of possible changes in the traffic profile. Therefore, partitioning the spectrum in a dynamic traffic environment is a challenging issue, which needs further research.

Some proposals [100,162] presented in the literature divide the spectrum into a number of partitions in an advance by assuming traffic demands in order to suppress the fragmentation in the network. Considering that the traffic condition can never be fully predicted, it may happen that a lightpath with a large number of required slots may not be supported by a partition; in this situation, a lightpath request tends to be rejected. This will increase call blocking in the network. To accommodate the lightpath with a large number of required slots, some partitions may need to be expanded. If the spectrum of a partition is not utilized, the partition should be shrunk automatically. Adjusting partition size to suit traffic changes is a challenging issue, which may be overcome by further research.

To suppress the bandwidth fragmentation in EONs, network operators should consider planning and designing phases. Mathematical modeling approaches [257], such as integer linear programming and mixed integer linear programming, are normally used to obtain the optimum output. When the network size increases, mathematical approaches are not tractable within a practical time. In that situation, heuristic approaches are one possible solution. In that case, optimality cannot be achieved, but computation time is practical. Therefore, designing efficient heuristic algorithms that minimize computational time while maximizing accuracy is a challenging issue.

As EONs have the ability to form super-wavelength channels that carry a huge amount of information in the order of Tb/s range, survivability against the failure of network components is an essential requirement. When current protection techniques are adopted, the possibility of increasing fragmented slots increases. To overcome this issue, some defragmentation proposals [276] have already been presented for path protected dynamic EONs; they consider the reallocation of backup paths as backup paths are used only if their corresponding primary paths fail. However, during the reallocation of backup paths, if failure occurs in primary path, the data cannot be protected. Therefore, we need defragmentation approaches for dynamic protected EONs, which can provide protection in any situation while managing the fragmentation effect concurrently.

The hitless defragmentation approach is considered the most effective one in handling fragmentation in EONs. As hop retuning does not support finely-granular systems, the push-pull retuning approach is promising to suppress bandwidth fragmentation. In push-pull retuning, the end-of-line problem is the main obstacle to defragmentation. Therefore, further study is needed to develop a scheme that avoids the end-of-line problem during push-pull retuning.

The non-hitless defragmentation approach may adopt the least common multiple (LCM) factor of required slots among all lightpath requests for handling the fragmentation issue in EONs. All the lightpath requests are allocated using the interval of the LCM factor. When lightpaths are torn down, they create fragmented slots. As the LCM interval is considered when spectrum slots are allocated, any new lightpath can easily be accommodated by the fragmented slots. This solution is useful for static traffic. In the case of dynamic traffic, it is difficult to estimate the LCM factor with existing approaches so further research is essential.

Some proposals on hitless defragmentation in 1+1 path protected EONs [276, 277] presented in the literature focuses only on defragmenting backup paths; the fragmentation caused by primary lightpaths is ignored. Therefore, we need to develop hitless defragmentation schemes for protected EONs that can handle the bandwidth fragmentation caused by both primary and backup paths.

Some pseudo partitioning defragmentation approaches [161, 174] can handle the fragmentation in EONs by considering that all lightpaths from the same source to the same destination follow the same route. The multiple end-to-end routing paths are not considered when distributing the traffic loads. If the multiple end-to-end routing paths are considered, the fragmentation effect in the network increases. Traffic fluctuation in multiple end-to-end routing paths must be investigated to understand the effect of fragmentation in pseudo partitioning defragmentation approaches, and further research is needed.

SDM technologies, such as bundle of single mode fibers and multi-core fibers, are introduced to overcome some physical barriers and enhance the overall capacity of optical transmission systems. Bandwidth fragmentation management during spectrum allocation in bundle of single mode fibers and multi-core fibers is a challenging issue, which needs further research.

13.6   Disaster management

Recovery of the network resources after large scale disasters such as—tsunami, hurricanes, floods, etc, is becoming increasingly important. The approach to disaster recovery is different from the approach to network failures such as fiber cuts or node failures. Network failures can be planned for by providing sufficient resources to deal with all network failures. However, there is no way to anticipate the impact of a disaster, and therefore no cost effective way to plan for full recovery from it. The recent development of SBVTs represents a new tool for handling disaster recovery. As SBVTs can play an important role in providing sufficient flexibility, elastic optical networks with SBVTs are an interesting research topic for disaster management.

Exercises

1. Why is amplification of multi-core fiber signals more challenging than that of single-core fiber signals?

2. Why does not hop retuning work in the system that is based on fine granularity?

3. Why is traffic grooming required for EONs, similar to WDM networks?

4. Describe some possible challenges for fragmentation management in EONs.

5. Why is managing physical layer impairments in EONs more challenging than that of WDM based optical networks?

Answers to Exercises

Chapter 1

Answer 1: (i) The amount of data that an optic fiber-based communication system transmits per unit time is far larger than the copper wire-based communication system. (ii) In an optic fiber-based communication system, optical fibers are capable of providing low power loss, which enables signals to be transmitted a longer distance than the copper wire-based communication system.

Answer 2: The key difference among frequency division multiplexing (FDM), time division multiplexing (TDM), and wavelength division multiplexing (WDM) is that FDM divides the bandwidth into smaller frequency ranges and each user transmits data simultaneously through a common channel, which is a type of transmission media that is used to transfer a message from one point to another, within their frequency range. TDM allocates a fixed time slot for each user to send signals through a common channel while WDM combines multiple light beams from several channels and combine them to a single light beam and sends through a fiber optic strand similar to FDM.

Answer 3: Dense wavelength division multiplexing (DWDM) technology is being deployed by several telecommunication companies for point-to-point communications. A W-channel based DWDM solution, (see Fig. 1.5), where a DWDM multiplexer (OMUX) combines W independent data streams, each on a unique wavelength, and sends them on a fiber, and a DWDM demultiplexer (ODMUX) at the fiber's receiving end separates these data streams into point-to-point links.

Answer 4: According to (1.2), we estimate wavelength spacing:

$$\Delta\lambda = \frac{(1520 \times 10^{-9})^2}{3 \times 10^8} \times 80 \times 10^9 = 6.16 \times 10^{-10} \text{ meter} = 0.6161 \text{ nm}$$

Answer 5: The numbers of channels for S band $= \frac{1530-1460}{0.8} = 88$. The numbers of channels for C band $= \frac{1565-1530}{0.8} = 44$. The numbers of channels for L band $= \frac{1625-1565}{0.8} = 75$.

Answer 6: The key advantages of SDH over PDH are: (i) the standardized optical interfaces make SDH more convenient for interconnection in lines, (ii) better network reliability, (iii) lower equipment investment, and (iv) better connectivity between different telecom carriers.

Answer 7:

   i   SONET path layer.

   ii   SONET line layer.

   iii   SONET section layer.

   iv   SONET section layer.

   v   SONET section layer.

Answer 8: Physical layer impairments can be classified into two categories, which are (i) linear impairments: attenuation and dispersion (ii) nonlinear impairment: self-phase modulation (SPM), four wave mixing (FWM), cross phase modulation (XPM), stimulated brillouin scattering (SBS), and stimulated Raman scattering (SRS).

Answer 9: According to (1.1), we estimate the transmission reach:

$$L = -\frac{10}{0.03}\log_{10}\frac{0.05}{0.2} = -333.333 \times (-0.602) = 200.69 \text{ km}.$$

Answer 10: A lightpath is a path that consists of optical fiber links between a source-destination pair. When a lightpath is established between a source-destination pair, the connection is totally optical and avoids throttling by intermediate electronic conversions and processing. The electronic conversions and processing at intermediate nodes increases additional network cost, leading to serving the performance bottleneck and restraining the provision of optical link bandwidth to end users. To establish a lightpath, intermediate network elements including node are not required to be designed for different bit rates and modulation formats.

Answer 11: Passive-star based optical networks are typically used in local area networks and metropolitan area networks as the signal power is split among various nodes. Therefore this type of network cannot be used in long distance communication. Whereas, wavelength-routed optical networks are used in wide area networks as they support wavelength reuse property and the signal power is not split among various nodes.

Answer 12: (i) Two lightpaths can use the same wavelength if they use a link disjoint path. (ii) If two lightpaths share a link, they can be assigned different wavelengths.

Answer 13: In Fig. 1.7, each lightpath uses the same wavelength on all hops in the end-to-end path due to wavelength continuity constraint property. The established lightpaths between source-destination pairs A-C and B-F use different wavelengths because they use the common fiber link 6-7. This property is known as the distinct channel constraint. The established lightpaths between source-destination pairs H-G and D-E use the same wavelength, which is already used by the lightpath A-C due to a wavelength reuse characteristic.

Answer 14: In the given network, there are 10 links. Each link has two wavelengths. The wavelength usage pattern for the network is given below.

Link	λ_1	λ_2
AB	Lightpath request AB	Lightpath request AD
BC	Lightpath request BC	Lightpath request AD
CD	Lightpath request CD	Lightpath request AD
AF	Lightpath request AC	Lightpath request AE
FE	Lightpath request FE	Lightpath request AE
ED	Lightpath request BD	Lightpath request ED
BF	Lightpath request BF	
EC	Lightpath request EC	
BE	Lightpath request BD	Lightpath request BE
FC	Lightpath request AC	Lightpath request FC

The given lightpath requests are AB, AC, AD, AE, AF, BC, BD, BE, BF, CD, EC, FC, ED, FD, and FE, which arrive in the network sequentially. Lightpath request AB is allocated on link AB using first fit wavelength assignment policy; wavelength λ_1 and route A-B are used. There are six possible routes for lightpath request AC, which are A-B-C, A-B-F-C, A-F-C, A-B-E-C, A-B-F-E-C, and A-F-E-C. As minimum hop routing is considered, either route A-B-C or route A-F-C is used. Route A-F-C is chosen randomly for lightpath establishment. The used routes of remaining lightpath requests AD, AE, AF, BC, BD, BE, BF, CD, EC, FC, ED, FD, and FE are A-B-C-D, A-F-E, A-F, B-C, B-E-D, B-E, B-F, C-D, E-C, F-C, E-D, F-E-D, and F-E, respectively.

Analysis: The lightpath requests AF and FD are not established due to unavailability of wavelength. Therefore, the blocking ratio in the network $= 2/15 = 0.133$.

Answer 15: In the given network, there are 10 links. Each link has two wavelengths. The wavelength usage pattern for the network is given below.

Link	λ_1	λ_2
AB	Lightpath request AB	Lightpath request AC
BC	Lightpath request BC	Lightpath request AC
CD	Lightpath request CD	
AF	Lightpath request AD	Lightpath request AF
FE	Lightpath request AD	Lightpath request FE
ED	Lightpath request AD	Lightpath request BD
BF	Lightpath request BF	
EC	Lightpath request EC	
BE	Lightpath request BE	Lightpath request BD
FC	Lightpath request FC	

The given lightpath requests are AB, AC, AD, AE, AF, BC, BD, BE, BF, CD, EC, FC, ED, FD, and FE, which arrive in the network sequentially. Lightpath request AB is allocated on link AB using first fit wavelength assignment policy; wavelength λ_1 and route A-B are used. There are six possible routes for lightpath request AC, which are A-B-C, A-B-F-C, A-F-C, A-B-E-C, A-B-F-E-C, and A-F-E-C. As minimum hop alternate path routing is considered, either route A-B-C or route A-F-C is used. As route A-F-C is chosen already in the previous question, route A-F-C is not used besides, route A-B-C chosen for lightpath establishment. The used routes of remaining lightpath requests AD, AE, AF, BC, BD, BE, BF, CD, EC, FC, ED, FD, and FE are A-F-E-D, A-B-E, A-F, B-C, B-E-D, B-E, B-F, C-D, E-C, F-C, E-D, F-E-D, and F-E, respectively,.

Analysis: Lightpath requests AE, ED, and FD are not established due to unavailability of wavelength. In this example, as alternate paths are used for lightpath requests AC, AD, and AE, instead of minimum hop routing, more spectrum resources are involved compared to minimum hop routing. Therefore, the blocking ratio in the network, which is $= 3/15 = 0.2$, is increased compared to the previous example.

Answer 16: DWDM based optical networks use ITU fixed spectrum grid of 100 GHz spacing. If the channels carry only low bandwidth, and no traffic can be transmitted in the large unused frequency gap, a large portion of the spectrum will be wasted. As a result, the resource utilization using DWDM based optical networks is suppressed.

Chapter 2

Answer 1: Bit rate is defined as the transmission of number of bits per second. Baud rate is defined as the number of signal units per second. The relationship between bit rate and baud rate is:

$$\text{Bit rate} = \text{Baud rate} \times \text{Number of bits per Baud.}$$

Answer 2: The conventional WDM network has its own limitations, as follows. (i) It follows ITU-T 100GHz or 50GHz fixed grid frequency spacing policy. Due to this large frequency spacing, a large portion of the frequency spectrum is wasted. (ii) It can provide bandwidth upto 100Gbps only, which is not sufficient for the future. To overcome these limitations, the elastic optical network (EON) is introduced, which provides flexibility in frequency spacing and offers bandwidth beyond 100Gbps.

Answer 3: OFDM is a special class of multi-carrier modulation (MCM) scheme that transmits a high-speed data stream by dividing it into a number of orthogonal channels, each of which carries relatively low-data rate compared to WDM systems, where a fixed channel spacing between two wavelengths is needed to eliminate crosstalk. EONs with OFDM technology allows the spectrum of individual subcarriers to overlap because of its orthogonality, which increases spectral efficiency (see Figs. 2.4 and 2.5).

Answer 4: The segmentation and aggregation properties are used to create sub and super wavelength channels, respectively, in EONs, which are explained in section 2.3.

Answer 5: OFDM splits the data stream into several sub-streams, which are sent in parallel on several subcarriers. Each subcarrier can be modulated based on bit rate requirement and transmission reach. The peak amplitude of any subcarrier's spectrum coincides with the zero point of other subcarriers' spectra, which means that when a subcarrier is sampled at its own peak, all other subcarriers cross at zero point. Thus, they do not interfere with each other and no guard band is required to separate them. Due to the orthogonality, OFDM achieves spectrum efficiency. Keeping the bandwidth requirement constant (the same data rate), a less robust modulation format, such as quadrature phase shift keying (QPSK), carries twice the number of bits per symbol than a more robust modulation format, such as binary phase shift keying (BPSK), which means the baud rate for QPSK is the half of BPSK. When the baud rate of each sub-stream is reduced, we save spectral bandwidth, and hence spectrum efficiency is increased. Therefore, if QPSK is used instead of BPSK, spectrum utilization is enhanced. The optical signal to noise ratio (OSNR) degradation using QPSK is larger than that of the BPSK. Therefore, the transmission reach using QPSK is shorter than BPSK.

Answer 6: The main technical issues that impair the performance of OFDM are frequency offset and timing jitter, which degrade the orthogonality of the subchannels.

Answer 7: The unique properties of EONs over WDM based optical networks are: (i) bandwidth segmentation, (ii) bandwidth aggregation, (iii) efficient accommodation of multiple data-rates, (iv) elastic variation of allocated resources, and (v) reach adaptation line rate.

Answer 8: Bandwidth-variable transponder (BVT) support high-speed transmission using spectrally efficient modulation formats, e.g., 16-quadrature amplitude modulation (16-QAM) is used for shorter distance lightpaths. Longer distance lightpaths are supported by using more robust but less efficient modulation formats, e.g., quadrature phase-shift keying (QPSK) or binary phase-shift keying (BPSK). Therefore, BVTs are able to trade spectral efficiency off against transmission reach.

Answer 9: We know that $f = c/\lambda$, $f_1 = c/\lambda_1$, $c = 3 \times 10^8$ m/s, $\lambda_1 = 1530$ nm
$f_1 = 3 \times 10^8/1530 \times 10^{-9} = 196.1$ THz
$f_2 = c/\lambda_2, c = 3 \times 10^8$ m/s, $\lambda_2 = 1565$ nm
$f_2 = 3 \times 10^8/1565 \times 10^{-9} = 191.7$ THz
$f_1 - f_2 = 196.1 - 191.7 = 4.4$ THz
Number of spectrum slots $= (f_1 - f_2)/(\Delta f) = 4.4$ THz / 12.5 GHz = 352.

Answer 10: The number of required slots $(F) = b/(s \times m)$, where b, s, and m represent bit rate, spectrum slot granularity, and modulation level, respectively. Given, $b = 100$ Gbps, $s = 12.5$ Gbps, $m = 1, 2, 3, 4$.

    i For lightpath request AB, $m = 4$ (16QAM is used as the transmission reach is 1000 km). Therefore, the number of required slots for lightpath request AB = 100 / (12.5 × 4) = 2 slots.

    ii For lightpath request AC, $m = 2$ (QPSK is used as the transmission reach is 3600 km). Therefore, the number of required slots for lightpath request AC = 100 / (12.5 × 2) = 4 slots.

    iii For lightpath request AD, $m = 1$ (BPSK is used as the transmission reach is 4600 km). Therefore, the number of required slots for lightpath request AD = 100 / (12.5 × 1) = 8 slots.

    iv For lightpath request AE, $m = 2$ (QPSK is used as the transmission reach is 3800 km). Therefore, the number of required slots for lightpath request AE = 100 / (12.5 × 2) = 4 slots.

v   For lightpath request AF, *m* = 4 (16QAM is used as the transmission reach is 1000 km). Therefore, the number of required slots for lightpath request AF = 100 / (12.5 × 4) = 2 slots.

vi   For lightpath request BC, *m* = 3 (8QAM is used as the transmission reach is 2700 km). Therefore, the number of required slots for lightpath request BC = 100 / (12.5 × 3) = 2.67 slots = 3 slots.

vii   For lightpath request BD, *m* = 2 (QPSK is used as the transmission reach is 3700 km). Therefore, the number of required slots for lightpath request BD = 100 / (12.5 × 2) = 4 slots.

viii   For lightpath request BE, *m* = 3 (8QAM is used as the transmission reach is 2800 km). Therefore, the number of required slots for lightpath request BE = 100 / (12.5 × 3) = 2.67 slots = 3 slots.

ix   For lightpath request BF, *m* = 1 (BPSK is used as the transmission reach is 4600 km). Therefore, the number of required slots for lightpath request BF = 100 / (12.5 × 1) = 8 slots.

x   For lightpath request CD, *m* = 4 (16QAM is used as the transmission reach is 1000 km). Therefore, the number of required slots for lightpath request CD = 100 / (12.5 × 4) = 2 slots.

xi   For lightpath request EC, *m* = 1 (BPSK is used as the transmission reach is 4900 km). Therefore, the number of required slots for lightpath request EC = 100 / (12.5 × 1) = 8 slots.

xii   For lightpath request FC, *m* = 3 (8QAM is used as the transmission reach is 2600 km). Therefore, the number of required slots for lightpath request FC = 100 / (12.5 × 3) = 2.67 slots = 3 slots.

xiii   For lightpath request ED, *m* = 4 (16QAM is used as the transmission reach is 1000 km). Therefore, the number of required slots for lightpath request ED = 100 / (12.5 × 4) = 2 slots.

xiv   For lightpath request FD, *m* = 2 (QPSK is used as the transmission reach is 3600 km). Therefore, the number of required slots for lightpath request FD = 100 / (12.5 × 2) = 4 slots.

xv   For lightpath request FE, *m* = 3 (8QAM is used as the transmission reach is 2800 km). Therefore, the number of required slots for lightpath request FE = 100 / (12.5 × 3) = 2.67 slots = 3 slots.

Chapter 3

Answer 1: Bandwidth-variable transponders (BVTs) are capable of providing several dynamic functionalities that can be programmed by a remote controller, such as variable coding and overhead, modulation format, symbol rate and spectral shaping. This enables to support multiple rates (e.g., from 10 Gb/s to 1 Tb/s) and to provide the optimum spectrum usage depending on the transmission reach. When a high-speed BVT is operated at lower than its maximum rate due to the required reach or impairments in the optical path, a part of the BVT capacity is wasted. Sliceable bandwidth-variable transponders (SBVTs) have been presented that offer improved flexibility, which is capable to allocate its capacity into one or several optical flows that are transmitted to one or several destinations. Figure 3.2 distinguishes BVT and SBVT functionalities.

Answer 2: The advantage of spectrum routing node architecture, compared to the broadcast-and-select architecture, is that the power loss is not dependent on the number of degrees.

Answer 3: The spectrum routing node architecture does not support spectrum defragmentation. Whereas, the switch and select node architecture with dynamic functionality supports defragmentation.

Answer 4: The following properties are considered to design a node that can be an appropriate candidate for a flexible node architecture.

  i Flexibility: the node must be designed in such way that it should provide channel flexibility, expansion flexibility, functional flexibility, switching flexibility, routing flexibility, and architectural flexibility.

  ii Colorless, directionless, and contentionless: to enable flexibility and re-configurability of the flexible optical nodes, flexible colorless, directionless, and contentionless (CDC) architectures with remote spectrum assignment are considered.

 iii Scalability: in order to provide the optical network scalability, it must be ensured that the node scalability is inherent in the node design.

  iv Resilience: optical node should be designed in such way that the network can be performed seamlessly in a very short period of time.

Answer 5: Please check section 3.2.5.

Answer 6: The architecture on demand (AoD) is considered one of the suitable node architectures for EONs to provide a cost-efficient solution. The number of spectrum selective switches and other processing devices is not fixed in AoD, but they can be determined based on the specific demand for those functionalities.

Thus, savings in the number of devices can balance the additional cost of the optical backplane, and hence AoD provides a cost-efficient solution. Furthermore, AoD provides considerable gains in terms of scalability and resiliency compared to other static architectures.

Chapter 4

Answer 1: The number of spectrum slots is estimated based on the users' bandwidth requirement. If the number of required slots is greater than one, the required slots are placed near to each other according to the spectrum contiguity constraint. If spectrum contiguity is violated, more slots are used as guardband to avoid interfaces with other lightpaths. As a result, the spectrum utilization is suppressed.

Answer 2: The major routing policies without considering elastic characteristics are fixed routing, alternate routing, and adaptive routing. The time complexity of the fixed routing policy is the lowest among all routing policies. However, its blocking probability is the highest. Adaptive routing provides the best performance in terms of blocking probability, but its time complexity is the highest. FAR offers a trade-off between time complexity and blocking probability.

Answer 3: Please check section 4.3.4.

Answer 4: In the given network, there are 10 links. Each link has five slots, which are slot 1, slot 2, slot 3, slot 4, and slot 5. The spectrum slot usage pattern for the network is given below.

Link	Slot 1	Slot 2	Slot 3	Slot 4	Slot 5
AB	Request AB	Request AB	Request AD	Request AD	
BC	Request BC	Request BC	Request AD	Request AD	
CD	Request CD	Request CD	Request AD	Request AD	
AF	Request AC	Request AC	Request AE	Request AE	
FE	Request FE	Request FE	Request AE	Request AE	
ED	Request BD	Request BD	Request ED	Request ED	
BF	Request BF	Request BF			
FC	Request AC	Request AC	Request FC	Request FC	
BE	Request BD	Request BD	Request BE	Request BE	
EC	Request EC	Request EC			

Lightpath request AB is allocated to slot 1 and slot 2 on link AB using the first fit spectrum allocation policy; route A-B is used. There are six possible routes for lightpath request AC, which are A-B-C, A-B-F-C, A-F-C, A-B-E-C, A-B-F-E-C, and A-F-E-C. As minimum hop routing is considered, either route A-B-C or route A-F-C is used. Route A-F-C is chosen randomly for lightpath establishment. The routes of remaining lightpath requests AD, AE, AF, BC, BD, BE, BF, CD, EC, FC, ED, FD, and FE are A-B-C-D, A-F-E, A-F, B-C, B-E-D, B-E, B-F, C-D, E-C, F-C, E-D, F-E-D, and F-E, respectively.

Analysis: The lightpath requests AF and FD are not established due to unavailability of required slots. Therefore, the blocking ratio in the network $= 2/15 = 0.133$.

Answer 5: In the given network, there are 10 links. Each link has five slots, which are slot 1, slot 2, slot 3, slot 4, and slot 5. The spectrum slot usage pattern for the network is given below.

Link	Slot 1	Slot 2	Slot 3	Slot 4	Slot 5
AB	Request AB	Request AB	Request AC	Request AC	
BC	Request BC	Request BC	Request AC	Request AC	
CD	Request CD	Request CD			
AF	Request AD	Request AD	Request AF	Request AF	
FE	Request AD	Request AD	Request FE	Request FE	
ED	Request AD	Request AD	Request BD	Request BD	
BF	Request BF	Request BF			
FC	Request FC	Request FC			
BE	Request BE	Request BE	Request BD	Request BD	
EC	Request EC	Request EC			

Lightpath request AB is allocated to slot 1 and slot 2 on link AB using the first fit spectrum allocation policy; route A-B is used. There are six possible routes for lightpath request AC, which are A-B-C, A-B-F-C, A-F-C, A-B-E-C, A-B-F-E-C, and A-F-E-C. As minimum hop alternate path routing is considered, either route A-B-C or route A-F-C is used. Route A-F-C is chosen already in the previous question so route A-F-C is not used besides; route A-B-C is chosen for lightpath establishment. The routes of remaining lightpath requests AD, AE, AF, BC, BD, BE, BF, CD, EC, FC, ED, FD, and FE are A-F-E-D, A-B-E, A-F, B-C, B-E-D, B-E, B-F, C-D, E-C, F-C, E-D, F-E-D, and F-E, respectively.

Analysis: Lightpath requests AE, ED, and FD are not established due to unavailability of required slots. In this example, as alternate paths are used for lightpath requests AC, AD, and AE, instead of minimum hop routing, more spectrum resources are involved compared to minimum hop routing. Therefore, the blocking ratio in the network is $= 3/15 = 0.2$, which is increased compared to the previous example.

Answer 6: In the given network, there are 10 links. Each link has five slots, which are slot 1, slot 2, slot 3, slot 4, and slot 5. The spectrum slot usage pattern for the network is given below:

Link	Slot 1	Slot 2	Slot 3	Slot 4	Slot 5
AB	Request AB	Request AB	Request AD	Request AD	
BC	Request BC	Request BC	Request AD	Request AD	
CD	Request CD	Request CD	Request AD	Request AD	
AF		Request AE	Request AE	Request AC	Request AC
FE		Request AE	Request AE	Request FE	Request FE
ED	Request BD	Request BD	Request ED	Request ED	
BF	Request BF	Request BF			
FC		Request FC	Request FC	Request AC	Request AC
BE	Request BD	Request BD		Request BE	Request BE
EC	Request EC	Request EC			

In the first-last fit policy, odd and even indexed lightpath requests are allocated using first fit and last fit, respectively. Lightpath request AB is allocated to slot 1 and slot 2 on link AB using the first fit spectrum allocation policy as it is an odd request; route A-B is used. There are six possible routes for lightpath request AC, which are A-B-C, A-B-F-C, A-F-C, A-B-E-C, A-B-F-E-C, and A-F-E-C. As minimum hop routing is considered, either route A-B-C or route A-F-C is used. Route A-F-C is chosen randomly for lightpath establishment. Lightpath request AC is allocated to slot 4 and slot 5 using the last fit policy as it is an even request. The routes of remaining lightpath requests AD, AE, AF, BC, BD, BE, BF, CD, EC, FC, ED, FD, and FE are A-B-C-D, A-F-E, A-F, B-C, B-E-D, B-E, B-F, C-D, E-C, F-C, E-D, F-E-D, and F-E, respectively.

Analysis: The lightpath requests AF, and FD are not established due to unavailability of required slots. Therefore, the blocking ratio in the network $= 2/15 = 0.133$.

Answer 7: In the given network, there are 10 links. Each link has five slots, which are slot 1, slot 2, slot 3, slot 4, and slot 5. The spectrum slot usage pattern for the network is given below.

Lightpath request AB is allocated to slot 1 and slot 2 on link AB using the first fit spectrum allocation policy. There are six possible routes for lightpath request AC, which are A-B-C, A-B-F-C, A-F-C, A-B-E-C, A-B-F-E-C, and A-F-E-C. As spectrum split routing is considered, lightpath request AC requires two slots; one slot uses route A-B-C and the other slot uses route A-F-C. The used routes of remaining lightpath requests AD, AE, AF, BC, BD, BE, BF, CD, EC, FC, ED, FD, and FE are A-B-C-D, A-F-E, A-F, B-C, B-E-D, B-E, B-F, C-D, E-C, F-C, E-D, F-E-D, and F-E, respectively.

Link	Slot 1	Slot 2	Slot 3	Slot 4	Slot 5
AB	Request AB	Request AB	Request AC	Request AD	Request AD
BC	Request BC	Request BC	Request AC	Request AD	Request AD
CD	Request CD	Request CD		Request AD	Request AD
AF	Request AE	Request AE	Request AC	Request AF	Request AF
FE	Request AE	Request AE	Request FE	Request FE	
ED	Request BD	Request BD	Request ED	Request ED	
BF	Request BF	Request BF			
FC	Request FC	Request FC	Request AC		
BE	Request BD	Request BD	Request BE	Request BE	
EC	Request EC	Request EC			

Analysis: The lightpath request FD is not established due to unavailability of required slots. Therefore, the blocking ratio in the network $= 1/15 = 0.066$.

Answer 8: In fixed spectrum allocation, both central frequency (CF) and assigned spectrum width remain static for ever. In semi-static spectrum allocation, the CF remains fixed, but the allocated spectrum width can vary in each time interval. In elastic spectrum allocation, both central frequency and spectrum width vary dynamically. Therefore, elastic spectrum allocation provides the best performance in terms of blocking ratio compared to fixed and semi-elastic spectrum allocation.

Answer 9: The performances of first fit spectrum allocation, in terms of both blocking ratio and time complexity, is better compared to other allocation policies. Therefore, first fit spectrum allocation is considered one of the most appropriate spectrum allocation approaches in the literature.

Answer 10: Please check Tables 4.1 and 4.3.

Chapter 5

Answer 1: Proactive and reactive fragmentation-aware RSA approaches are not mutually exclusive. One can make some approaches considering both proactive and reactive fragmentation-aware RSA approaches. In the proactive fragmentation-aware RSA approaches, necessary precautions are taken to avoid fragmentation before the establishment of a lightpath. However, no action is taken for in-service lightpaths. Whereas in the case of reactive fragmentation-aware RSA approaches, a necessary action is taken for in-service lightpaths in order to suppress the fragmentation effect.

Answer 2: The protection techniques use backup paths to carry optical signals after fault occurrence, and these backup paths are computed prior to fault occurrence, but they are reconfigured after fault occurrence. Whereas, in restoration, backup paths are computed dynamically based on the latest network information after fault occurrence.

As protection techniques discover backup paths prior to fault occurrence, they offer faster recovery than the restoration techniques. As restoration techniques discover the backup paths dynamically based on the latest network information after fault occurrence, they do not reserve any resource prior to fault occurrence, and hence they provide better resource utilization than protection techniques.

Answer 3: EONs use sliceable bandwidth variable transponders (SBVTs) that can be reconfigured in such way that some network elements can move into sleep mode when the network traffic is below a certain threshold and avoid electrical processing, which reduces energy consumption. This facility is not provided by traditional transponders that are used for WDM based optical networks.

Answer 4: SBVTs allow the reuse of hardware and optical spectrum by transmitting data to multiple destinations. They provide point to multiple point transmission where the traffic rate to each destination and the number of destinations can be freely set to satisfy the request. On the other hand, the non-sliceable transponder requires at least one interface for each destination, which increases networking cost.

Answer 5: A higher-level modulation format is suitable for longer transmission reach, and a lower-level modulation format is suitable to shorter transmission reach, which is depicted in Fig. 5.2.

Answer 6: Fragmentation of spectrum creates non-aligned and non-continuous spectrum slots that are difficult to be utilized by future requests. As a result, the blocking of requests in the network is increased.

Chapter 6

Answer 1:

 i According to (6.1), the fragmentation effect for the link mentioned in Fig. 6.7 using the external fragmentation metric is estimated in the following. $A = \max\{1,4,3,4\} = 4$, $B = 1+4+3+4 = 12$
$\Theta = 1 - (4/12) = 0.67$

 ii According to (6.2), the fragmentation effect for the link mentioned in Fig. 6.7 using the entropy-based fragmentation metric is estimated in the following.

$$S = 17, I = \{1,2,3,4\}, f_1 = 1, f_2 = 4, f_3 = 3, f_4 = 4.$$

$$\tau = \frac{f_1}{S}\ln\left(\frac{S}{f_1}\right) + \frac{f_2}{S}\ln\left(\frac{S}{f_2}\right) + \frac{f_3}{S}\ln\left(\frac{S}{f_3}\right) + \frac{f_4}{S}\ln\left(\frac{S}{f_4}\right)$$

$$\tau = \frac{1}{17}\ln\left(\frac{17}{1}\right) + \frac{4}{17}\ln\left(\frac{17}{4}\right) + \frac{3}{17}\ln\left(\frac{17}{3}\right) + \frac{4}{17}\ln\left(\frac{17}{4}\right) = 1.154$$

iii According to (6.3), the fragmentation effect for the link mentioned in Fig. 6.7 using the access blocking probability metric is estimated in the following.

$G_1 = 3$, $G_2 = 4$, $I = \{1,2,3,4\}$, $f_1 = 1$, $f_2 = 4$, $f_3 = 3$, $f_4 = 4$,

$B = 1 + 4 + 3 + 4 = 12$

$$\sigma = 1 - \frac{\text{DIV}(1,3) + \text{DIV}(1,4) + \text{DIV}(4,3) + \text{DIV}(4,4) + \text{DIV}(3,3) + \text{DIV}(3,4) + \text{DIV}(4,3) + \text{DIV}(4,4)}{\text{DIV}(12,3) + \text{DIV}(12,4)}$$

$$\sigma = 1 - \frac{0+0+1+1+1+0+1+1}{4+3} = 0.28$$

Answer 2: We consider six lightpath requests, which are AB $(d = 1)$, AC $(d = 2)$, AD $(d = 3)$, BC $(d = 4)$, BD $(d = 5)$, and DC $(d = 6)$. The routes of AB, AC, AD, BC, BD, and DC are A-B, A-B-C, A-B-D, B-C, B-D, and D-C, respectively. Lightpath requests are established using the first fit allocation policy and after allocation the spectrum condition in the network is shown in Fig. 13.2. According to section 6.2.2, the fragmentation effect for the network (see Fig. 13.2) is estimated in the following.

$Z = 15$, $|D| = 6$, $|K_d| = 1$ for all $d \in D$, $w_{d1} = 1/6$ is set.

We obtain $\gamma_{11} = 3$, $\gamma_{21} = 6$, $\gamma_{31} = 6$, $\gamma_{41} = 3$, $\gamma_{51} = 4$, and $\gamma_{61} = 4$.

As $\Psi_{dk} = \gamma_{dk}/Z$, we estimate

$\Psi_{11} = 3/15$, $\Psi_{21} = 6/15$, $\Psi_{31} = 6/15$, $\Psi_{41} = 3/15$, $\Psi_{51} = 4/15$, and $\Psi_{61} = 4/15$.

$\phi = \sum_{d \in D} \sum_{k \in k_d} w_{dk} \cdot \Psi_{dk}$

$\phi = 1/6 \times 3/15 + 1/6 \times 6/15 + 1/6 \times 6/15 + 1/6 \times 3/15 + 1/6 \times 4/15 + 1/6 \times 4/15 = 13/45 = 0.29$

Fragmentation, $\chi = 1 - \phi = 1 - 0.29 = 0.71$

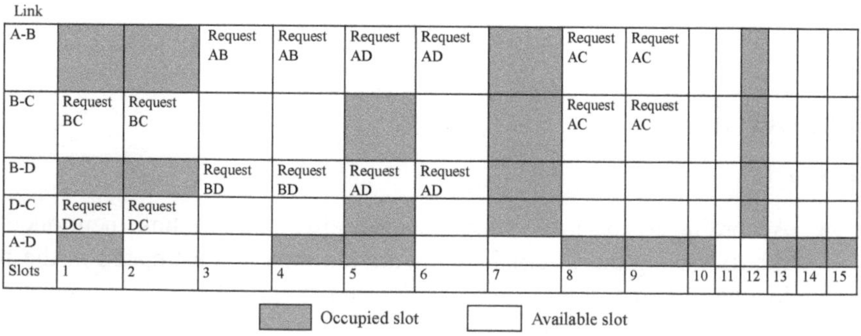

Link															
A-B			Request AB	Request AB	Request AD	Request AD		Request AC	Request AC						
B-C	Request BC	Request BC						Request AC	Request AC						
B-D			Request BD	Request BD	Request AD	Request AD									
D-C	Request DC	Request DC													
A-D															
Slots	1	2	3	4	5	6	7	8	9	10	11	12	13	14	15

Occupied slot ▢ Available slot

Figure 13.2: Spectrum allocation using first fit.

Answer 3: We consider six lightpath requests, which are AB $(d = 1)$, AC $(d = 2)$, AD $(d = 3)$, BC $(d = 4)$, BD $(d = 5)$, and DC $(d = 6)$. The routes of AB, AC, AD, BC, BD, and DC are A-B, A-B-C, A-B-D, B-C, B-D, and D-C, respectively. Lightpath requests are established using the first-last fit allocation policy and after allocation the spectrum condition in the network is shown in Fig. 13.3. According to section 6.2.2, the fragmentation effect for the network (see Fig. 13.3) is estimated in the following.

$Z = 15$, $|D| = 6$, $|K_d| = 1$ for all $d \in D$, $w_{d1} = 1/6$ is set.

We obtain $\gamma_{11} = 4$, $\gamma_{21} = 4$, $\gamma_{31} = 8$, $\gamma_{41} = 4$, $\gamma_{51} = 4$, and $\gamma_{61} = 4$.

As $\Psi_{dk} = \gamma_{dk}/Z$, we estimate $\Psi_{11} = 4/15$, $\Psi_{21} = 4/15$, $\Psi_{31} = 8/15$, $\Psi_{41} = 4/15$, $\Psi_{51} = 4/15$, and $\Psi_{61} = 4/15$.

$\phi = \sum_{d \in D} \sum_{k \in k_d} w_{dk} \cdot \Psi_{dk}$

$\phi = 1/6 \times 4/15 + 1/6 \times 4/15 + 1/6 \times 8/15 + 1/6 \times 4/15 + 1/6 \times 4/15 + 1/6 \times 4/15 = 14/45 = 0.31$

Fragmentation $(\chi) = 1 - \phi = 1 - 0.31 = 0.69$

Analysis: The first-last fit spectrum allocation policy suppresses the fragmentation compared to the first fit spectrum allocation policy.

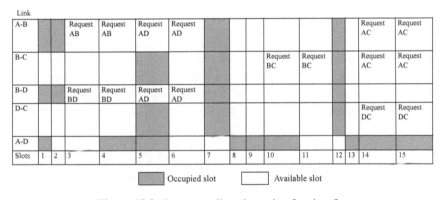

Figure 13.3: Spectrum allocation using first-last fit.

Chapter 7

Answer 1: The pseudo partitioning technique divides the entire spectrum into two partitions. On the other hand, in dedicated partitioning, the number of partitions must be greater than two. When the number of partitions increases, the system fails to offer statistical multiplexing gain. Due to the lack of statistical multiplexing gain, blocking of lightpath requests increases. When the number of partitions is high, the number of spectrum slots decreases in each partition, and hence the possibility of blocking of lightpath requests with high bandwidth requirement increases. As the number of partitions using the pseudo partitioning technique is less than that of the dedicated partitioning technique, the performance in terms of blocking ratio is better using the pseudo partitioning technique than that of the dedicated partitioning technique.

Answer 2: If the number of channels is 120 and offered traffic is 100 [Erlang], the blocking probability of the fiber link according to (7.1) is 0.0057.

Answer 3: Dividing 120 channels among 10 partitions and splitting the traffic among the partitions (12 channels with offered traffic volume of 10 [Erlang]), the blocking probability for each partition according to (7.1) is 0.1197.

Answer 4: The separation of disjoint and non-disjoint requests have a better impact in terms of fragmentation in the partitioning approach, as it provides a higher number of aligned available slots compared to the approach without the separation of requests.

Answer 5: The first-last fit spectrum allocation is more suitable for pseudo partitioning in terms of blocking ratio over the first fit policy, as it provides more number of contiguous available slots compared to the first fit policy.

Answer 6: The spectrum split routing relaxes the spectrum contiguity constraint. Using the spectrum split routing, traffic between a source-destination pair is transmitted through multiple routes to suppress fragmentation. When multiple routes are used, a larger number of slots are required as gardband in order to avoid interference with other lightpaths, which suppresses the spectrum utilization.

Answer 7: The average slot utilization is estimated according to (7.2).
$v = 100$, $|E| = 20$, $|F| = 10$, and average number of hops = 4
$U(v) = \frac{100 \times 4}{10 \times 20} = 2$

Answer 8: Confidence intervals are obtained at a confidence level, such as 95%, selected by users, which means that if the same population is sampled on numerous occasions and interval estimations are made on each occasion. The resulting intervals would contain the true population parameter in approximately 95% of the cases.

The smaller the margin of error, the more precisely we can pin point the population mean.

Chapter 8

Answer 1: The non-defragmentation approaches usually take proactive actions to suppress fragmentation. However, when a lightpath is torn down, they typically do not take any action. As a result, the torn down lightpath may creates fragmented slots. On the other hand, the defragmentation approaches retune the spectrum after a certain interval. Therefore, the possibility of filling the fragmented slots using the defragmentation approaches is higher than the non-defragmentation approaches.

Answer 2: The hop retuning approaches allow spectrum jump to fill-up the fragmented slots by retuning spectrum from any frequency to other frequency without causing any traffic disturbance. Any other approach does not allow spectrum jump from any frequency to other frequency. As the hop retuning approaches allow spectrum jump from any frequency to other frequency, it provides the best performance, in terms of suppressing fragmentation, theoretically compared to other defragmentation approaches. Note that, as the hop retuning technology demands sensitive spectrum sensing, it is difficult to implement in finely granular systems.

Answer 3: An end-of-line situation is defined by a situation where a lightpath cannot be retuned to fill in a gap left by an expired lightpath due to the interference of another lightpath preventing it from being moved further. When an end-of-line situation occurs, the retuning of a lightpath, if started, is stopped. As a result, it makes inefficient the defragmentation process.

Answer 4: The push-pull retuning time for a lightpath is estimated by (8.1).
$s = 10.5$ nm, $\alpha = 1$ ms, and $\beta = 10$ ms
Push-pull retuning time $(t_{\text{retuning}}) = 7 \times 1 + 10 = 17$ ms

Answer 5: Proactive defragmentation approaches are applied without waiting for a new lightpath request. On the other hand, reactive defragmentation approaches are normally triggered when a new lightpath request arrives in the network.

Answer 6: Please check section 8.1.2.4.

Answer 7: Please check the last paragraph of section 8.1.2.1.

Answer 8: Before tearing down of lightpaths AC and FE from the network, the spectrum slot usage pattern is the same as Answer 4 of Chapter 4. After tearing down of lightpaths AC and FE from the network, the spectrum slot usage pattern is given below.

Link	Slot 1	Slot 2	Slot 3	Slot 4	Slot 5
AB	Request AB	Request AB	Request AD	Request AD	
BC	Request BC	Request BC	Request AD	Request AD	
CD	Request CD	Request CD	Request AD	Request AD	
AF			Request AE	Request AE	
FE			Request AE	Request AE	
ED	Request BD	Request BD	Request ED	Request ED	
BF	Request BF	Request BF			
FC			Request FC	Request FC	
BE	Request BD	Request BD	Request BE	Request BE	
EC	Request EC	Request EC			

(i) After lightpaths AC and FE are tore down and before applying push-pull retuning, the contiguous-aligned available slot ratio in the network for active allocated lightpaths, which are AB ($d=1$), AD ($d=2$), AE ($d=3$), BC ($d=4$), BD ($d=5$), BE ($d=6$), BF ($d=7$), CD ($d=8$), EC ($d=9$), FC ($d=10$), and ED ($d=11$), is estimated in the following according to section 6.2.2.

$Z = 5, |D| = 11, |K_d| = 1$, for all $d \in D$, $w_{d1} = 1/11$ is set.

We obtain $\gamma_{11} = 1$, $\gamma_{21} = 3$, $\gamma_{31} = 4$, $\gamma_{41} = 1$, $\gamma_{51} = 2$, $\gamma_{61} = 1$, $\gamma_{71} = 3$, $\gamma_{81} = 1$, $\gamma_{91} = 3$, $\gamma_{101} = 2$, and $\gamma_{111} = 1$.

As $\Psi_{dk} = \gamma_{dk}/Z$, $\Psi_{11} = 1/5$, $\Psi_{21} = 3/5$, $\Psi_{31} = 4/5$, $\Psi_{41} = 1/5$, $\Psi_{51} = 2/5$, $\Psi_{61} = 1/5$, $\Psi_{71} = 3/5$, $\Psi_{81} = 1/5$, $\Psi_{91} = 3/5$, $\Psi_{101} = 2/5$, and $\Psi_{111} = 1/5$.

$\phi = \sum_{d \in D} \sum_{k \in k_d} w_{dk} \cdot \Psi_{dk}$

$\phi = 1/11 \times 1/5 + 1/11 \times 3/5 + 1/11 \times 4/5 + 1/11 \times 1/5 + 1/11 \times 2/5 + 1/11 \times 1/5 + 1/11 \times 3/5 + 1/11 \times 1/5 + 1/11 \times 3/5 + 1/11 \times 2/5 + 1/11 \times 1/5$

$\phi = 22/55 = 0.400$

Link	Slot 1	Slot 2	Slot 3	Slot 4	Slot 5
AB	Request AB	Request AB	Request AD	Request AD	
BC	Request BC	Request BC	Request AD	Request AD	
CD	Request CD	Request CD	Request AD	Request AD	
AF	Request AE	Request AE			
FE	Request AE	Request AE			
ED	Request BD	Request BD	Request ED	Request ED	
BF	Request BF	Request BF			
FC	Request FC	Request FC			
BE	Request BD	Request BD	Request BE	Request BE	
EC	Request EC	Request EC			

(ii) After push-pull retuning, the spectrum slot usage pattern is given below. The contiguous-aligned available slot ratio in the network is estimated in the following according to section 6.2.2.

We obtain $\gamma_{11} = 1$, $\gamma_{21} = 3$, $\gamma_{31} = 6$, $\gamma_{41} = 1$, $\gamma_{51} = 2$, $\gamma_{61} = 1$, $\gamma_{71} = 3$, $\gamma_{81} = 1$, $\gamma_{91} = 3$, $\gamma_{101} = 3$, and $\gamma_{111} = 1$.

As $\Psi_{dk} = \gamma_{dk}/Z$, $\Psi_{11} = 1/5$, $\Psi_{21} = 3/5$, $\Psi_{31} = 6/5$, $\Psi_{41} = 1/5$, $\Psi_{51} = 2/5$, $\Psi_{61} = 1/5$, $\Psi_{71} = 3/5$, $\Psi_{81} = 1/5$, $\Psi_{91} = 3/5$, $\Psi_{101} = 3/5$, and $\Psi_{111} = 1/5$.

$\phi = \sum_{d \in D} \sum_{k \in k_d} w_{dk} \cdot \Psi_{dk}$

$\phi = 1/11 \times 1/5 + 1/11 \times 3/5 + 1/11 \times 6/5 + 1/11 \times 1/5 + 1/11 \times 2/5 + 1/11 \times 1/5 + 1/11 \times 3/5 + 1/11 \times 1/5 + 1/11 \times 3/5 + 1/11 \times 3/5 + 1/11 \times 1/5$

$\phi = 25/55 = 0.454$

Analysis: The above results suggest that the push-pull retuning operation suppresses fragmentation in the network, and hence the contiguous-aligned available slot ratio in the network is improved.

Chapter 9

Answer 1: During the exchanging between primary and backup paths or reallocation of backup spectrum in 1+1 protected EONs, if any failure occurs in the primary path, data can not be recovered. Therefore, the defragmentation scheme using path exchanging provides quasi-1+1 protection, not 100% protection.

Answer 2: The defragmentation scheme using path exchanging allows it to exchange the path function of the 1+1 protection with the primary toggling to the backup state, while the backup becomes the primary. As a result, it avoids end-of-line conditions that are typically encountered in the push-pull retuning approach. Thus, the defragmentation scheme using path exchanging provides a better performance in terms of traffic admissibility than the push-pull retuning approach.

Answer 3: To execute the defragmentation scheme using path exchanging operations, it is assumed that no traffic disturbance will be triggered in the primary path during either exchanging between primary and backup paths or reallocation of the backup spectrum. The period of release during the reallocation process is short enough to support the 1+1 protection at almost every moment.

Answer 4: From the spectrum condition, it is observed that the network consists of 12 links, and each link has four spectrum slots. The traffic load and the average number of requested slots for each source-destination pair is the same over all the source-destination pairs.

(i) Before performing the defragmentation scheme using path exchanging, the contiguous-aligned available slot ratio in the network is estimated as follows.
$Z = 4$, $|D| = 7$, $|K_d| = 2$, for all $d \in D$, $w_{dk} = 1/14$ is set.
We get $\gamma_{11} = 2$, $\gamma_{12} = 2$, $\gamma_{21} = 2$, $\gamma_{22} = 2$, $\gamma_{31} = 2$, $\gamma_{32} = 2$, $\gamma_{41} = 3$, $\gamma_{42} = 0$, $\gamma_{51} = 3$, $\gamma_{52} = 4$, $\gamma_{61} = 3$, $\gamma_{62} = 2$, $\gamma_{71} = 6$ and $\gamma_{72} = 0$.
As $\Psi_{dk} = \gamma_{dk}/Z$, $\Psi_{11} = 2/4$, $\Psi_{12} = 2/4$, $\Psi_{21} = 2/4$, $\Psi_{22} = 2/4$, $\Psi_{31} = 2/4$,

$\Psi_{32} = 2/4$, $\Psi_{41} = 3/4$, $\Psi_{42} = 0/4$, $\Psi_{51} = 3/4$, $\Psi_{52} = 4/4$, $\Psi_{61} = 3/4$, $\Psi_{62} = 2/4$, $\Psi_{71} = 6/4$, and $\Psi_{72} = 0/4$.

$\phi = \sum_{d \in D} \sum_{k \in k_d} w_{dk} \cdot \Psi_{dk} = 1/14 \times 2/4 + 1/14 \times 2/4 + 1/14 \times 2/4 + 1/14 \times 2/4 + 1/14 \times 2/4 + 1/14 \times 2/4 + 1/14 \times 3/4 + 1/14 \times 0/4 + 1/14 \times 3/4 + 1/14 \times 4/4 + 1/14 \times 3/4 + 1/14 \times 2/4 + 1/14 \times 6/4 + 1/14 \times 0/4 = 33/56 = 0.589$.

(ii) After performing the defragmentation scheme using path exchanging, the spectrum condition in the network is given below. We obtain $\gamma_{11} = 2$, $\gamma_{12} = 2$,

Link	Slot 1	Slot 2	Slot 3	Slot 4
AB	Primary lightpath 1	Primary lightpath 7		
BC	Primary lightpath 2	Primary lightpath 7		
CD	Primary lightpath 3	Primary lightpath 7		
DE	Primary lightpath 4			
EF	Primary lightpath 5			
FA	Primary lightpath 6			
AG	Backup lightpath 1	Backup lightpath 6	Backup lightpath 7	
BG	Backup lightpath 1	Backup lightpath 2		
CG	Backup lightpath 3	Backup lightpath 2		
DG	Backup lightpath 3	Backup lightpath 4	Backup lightpath 7	
EG	Backup lightpath 5	Backup lightpath 4		
FG	Backup lightpath 5	Backup lightpath 6		

$\gamma_{21} = 2$, $\gamma_{22} = 4$, $\gamma_{31} = 2$, $\gamma_{32} = 2$, $\gamma_{41} = 3$, $\gamma_{42} = 2$, $\gamma_{51} = 3$, $\gamma_{52} = 4$, $\gamma_{61} = 3$, $\gamma_{62} = 2$, $\gamma_{71} = 6$, and $\gamma_{72} = 2$.

As $\Psi_{dk} = \gamma_{dk}/Z$, $\Psi_{11} = 2/4$, $\Psi_{12} = 2/4$, $\Psi_{21} = 2/4$, $\Psi_{22} = 4/4$, $\Psi_{31} = 2/4$, $\Psi_{32} = 2/4$, $\Psi_{41} = 3/4$, $\Psi_{42} = 2/4$, $\Psi_{51} = 3/4$, $\Psi_{52} = 4/4$, $\Psi_{61} = 3/4$, $\Psi_{62} = 2/4$, $\Psi_{71} = 6/4$, and $\Psi_{72} = 2/4$.

$\phi = \sum_{d \in D} \sum_{k \in k_d} w_{dk} \cdot \Psi_{dk} = 1/14 \times 2/4 + 1/14 \times 2/4 + 1/14 \times 2/4 + 1/14 \times 4/4 + 1/14 \times 2/4 + 1/14 \times 2/4 + 1/14 \times 3/4 + 1/14 \times 2/4 + 1/14 \times 3/4 + 1/14 \times 4/4 + 1/14 \times 3/4 + 1/14 \times 2/4 + 1/14 \times 6/4 + 1/14 \times 2/4 = 39/56 = 0.696$.

Observation: The defragmentation scheme using path exchanging improves the contiguous-aligned available slot ratio in the network.

Answer 5: From the spectrum condition, it is observed that the network consists of 12 links, and each link has four spectrum slots. The traffic load and the average number of requested slots for each source-destination pair is the same over all the source-destination pairs.

(i) Before performing the push-pull returning, the contiguous-aligned available slot ratio in the network is estimated as follows.

We obtain $\gamma_{11} = 2$, $\gamma_{12} = 2$, $\gamma_{21} = 1$, $\gamma_{22} = 2$, $\gamma_{31} = 2$, $\gamma_{32} = 2$, $\gamma_{41} = 2$, $\gamma_{42} = 0$,

$\gamma_{51} = 2$, $\gamma_{52} = 4$, $\gamma_{61} = 3$, $\gamma_{62} = 2$, $\gamma_{71} = 3$, and $\gamma_{72} = 0$.
As $\Psi_{dk} = \gamma_{dk}/Z$, $\Psi_{11} = 2/4$, $\Psi_{12} = 2/4$, $\Psi_{21} = 1/4$, $\Psi_{22} = 2/4$, $\Psi_{31} = 2/4$, $\Psi_{32} = 2/4$, $\Psi_{41} = 2/4$, $\Psi_{42} = 0/4$, $\Psi_{51} = 2/4$, $\Psi_{52} = 4/4$, $\Psi_{61} = 3/4$, $\Psi_{62} = 2/4$, $\Psi_{71} = 3/4$, and $\Psi_{72} = 0/4$.

$\phi = \sum_{d \varepsilon D} \sum_{k \varepsilon k_d} w_{dk} \cdot \Psi_{dk} = 1/14 \times 2/4 + 1/14 \times 2/4 + 1/14 \times 1/4 + 1/14 \times 2/4 + 1/14 \times 2/4 + 1/14 \times 2/4 + 1/14 \times 2/4 + 1/14 \times 0/4 + 1/14 \times 2/4 + 1/14 \times 4/4 + 1/14 \times 3/4 + 1/14 \times 2/4 + 1/14 \times 3/4 + 1/14 \times 0/4 = 27/56 = 0.482$

(ii) After performing push-pull retuning, the spectrum condition is given below. We obtain $\gamma_{11} = 2$, $\gamma_{12} = 2$, $\gamma_{21} = 1$, $\gamma_{22} = 2$, $\gamma_{31} = 2$, $\gamma_{32} = 2$, $\gamma_{41} = 3$, $\gamma_{42} = 0$,

Link	Slot 1	Slot 2	Slot 3	Slot 4
AB	Primary lightpath 1	Primary lightpath 7		
BC		Primary lightpath 7	Primary lightpath 2	
CD	Backup lightpath 3	Primary lightpath 7		
DE	Primary lightpath 4			
EF	Primary lightpath 5			
FA	Primary lightpath 6			
AG	Backup lightpath 1	Backup lightpath 6	Backup lightpath 7	
BG	Backup lightpath 1	Backup lightpath 2		
CG		Backup lightpath 2		Primary lightpath 3
DG		Backup lightpath 4	Backup lightpath 7	Primary lightpath 3
EG	Backup lightpath 5	Backup lightpath 4		
FG	Backup lightpath 5	Backup lightpath 6		

$\gamma_{51} = 3$, $\gamma_{52} = 4$, $\gamma_{61} = 3$, $\gamma_{62} = 2$, $\gamma_{71} = 3$, and $\gamma_{72} = 0$.
As $\Psi_{dk} = \gamma_{dk}/Z$, $\Psi_{11} = 2/4$, $\Psi_{12} = 2/4$, $\Psi_{21} = 1/4$, $\Psi_{22} = 2/4$, $\Psi_{31} = 2/4$, $\Psi_{32} = 2/4$, $\Psi_{41} = 3/4$, $\Psi_{42} = 0/4$, $\Psi_{51} = 3/4$, $\Psi_{52} = 4/4$, $\Psi_{61} = 3/4$, $\Psi_{62} = 2/4$, $\Psi_{71} = 3/4$, and $\Psi_{72} = 0/4$.

$\phi = \sum_{d \varepsilon D} \sum_{k \varepsilon k_d} w_{dk} \cdot \Psi_{dk} = 1/14 \times 2/4 + 1/14 \times 2/4 + 1/14 \times 1/4 + 1/14 \times 2/4 + 1/14 \times 2/4 + 1/14 \times 3/4 + 1/14 \times 0/4 + 1/14 \times 3/4 + 1/14 \times 4/4 + 1/14 \times 3/4 + 1/14 \times 2/4 + 1/14 \times 3/4 + 1/14 \times 0/4 = 29/56 = 0.517$

Observation: The defragmentation using push-pull retuning improves the contiguous-aligned available slot ratio in the network.

Answer 6: The defragmentation scheme using path exchanging allows us to exchange the path function of the 1+1 protection with the primary toggling to the backup state, while the backup becomes the primary. It suppresses fragmentation caused by both primary and backup paths. The defragmentation using only backup path reallocation suppresses fragmentation caused by only backup paths; the fragmentation caused by primary lightpaths are overlooked. When fragmentation is suppressed, the traffic admissibility in the network is improved. Therefore, the defragmentation scheme using path exchanging provides better performance in terms of traffic admissibility than only backup path reallocation.

Chapter 10

Answer 1: Network function virtulization (NFV) implements network functions, such as routing, security, QoS, load balancing, etc, in software instead of doing it on specialised hardware. NFV is basically picking any network function and implementing it as a service on a virtual machine on traditional server hardware with minimalistic general purpose accelerators. SDN seeks to separate network control functions from network forwarding functions, while NFV seeks abstract network forwarding and other networking functions from the hardware on which it runs.

Answer 2: Openflow is a protocol, which gives standard specifications for communication between the SDN controller and network equipments, specially switches. It allows routing decisions to be taken by SDN controllers and let forwarding rules, and security rules being pushed on switches in underlying networks.

Answer 3: According to the SD-EON architecture, there are three main components, which are the controllers, the forwarding devices and the communication protocols between control plane and data plane.

Answer 4: Please check the third paragraph of section 10.1 and Fig. 10.2.

Answer 5: The major difference between traditional SDN and SD-EON architecture is due to the network elements that form the data plane. Typically, bandwidth-variable transponders and switching components are used to build the data plane of SD-EON. On the other hand, normal switches are used in the traditional SDN. The resource in the SD-EON control plane is spectrum slots, which is another difference between SD-EON and SDN. Furthermore, when routes are computed, the control plane in SD-EON considers several aspects, such as modulation formats and number of required spectrum slots, which are not considered in SDN.

Answer 6: Please check the last two paragraphs of section 10.2.

Answer 7: The SD-EON is intended to facilitate several unique properties, which include bandwidth segmentation, bandwidth aggregation, efficient accommodation of multiple data rates, elastic variation of allocated resources, and reach-adaptable line rate. The existing OpenFlow protocols are unable to support these unique properties. Therefore, they need to be updated for SD-EONs.

Chapter 11

Answer 1: Integer linear programming (ILP) is a subset of the broader field of linear programming (LP). Both are seeking optimal values (either in the minimization or maximization sense) of an objective function with a set of decision variables. In an ILP problem, decision variables take only integer values, whereas, in an LP problem, decision variables can be real numbers. In general, an ILP problem takes more time than an LP one.

Answer 2: The following steps are used to solve an LP problem using the corner point method. They are (i) find the feasible region of the LP problem, (ii) find the co-ordinates of each corner point of the feasible region; these co-ordinates can be obtained by solving the multiple equations provided by constraints, (iii) at each corner point, compute the value of the objective function, (iv) identify the corner point at which the value of the objective function is maximum or minimum depending on the LP problem.

Answer 3: The optimal solution is obtained at $x_1 = 10/3$ and $x_2 = 4/3$; the optimal value for objective function is $\frac{160}{3}$.

Answer 4: The optimal solution is obtained at $x_1 = 2$ and $x_2 = 2$; the optimal value for objective function is 50.

Answer 5:

$$\text{min} \quad 500x_{12} + 300x_{13} + 100x_{23} + 200x_{24} + 300x_{34} \quad (13.1\text{a})$$
$$\text{s.t.} \quad x_{12} + x_{13} = 1 \quad (13.1\text{b})$$
$$x_{12} - x_{23} - x_{24} = 0 \quad (13.1\text{c})$$
$$x_{13} + x_{23} - x_{34} = 0 \quad (13.1\text{d})$$
$$0 \le x_{12} \le 1 \quad (13.1\text{e})$$
$$0 \le x_{13} \le 1 \quad (13.1\text{f})$$
$$0 \le x_{23} \le 1 \quad (13.1\text{g})$$
$$0 \le x_{24} \le 1 \quad (13.1\text{h})$$
$$0 \le x_{34} \le 1 \quad (13.1\text{i})$$

The optimal solution is obtained at $x_{12} = 0, x_{13} = 1, x_{23} = 0, x_{24} = 0$, and $x_{34} = 1$; the optimal value for objective function is 600. As a result, the shortest path is determined as: node 1 $\rightarrow$ node 3 $\rightarrow$ node 4.

Chapter 12

Answer 1:

We select $n_0 = 1$ and $c = 25, \forall n \geq n_0$.

$$5n^2 - 3n + 20 \leq 5n^2 + 20n^2 \text{ as } n \geq 1$$
$$= 25n^2$$
$$= cn^2$$

Thus, by definition of big O, $5n^2 - 3n + 20$ is in $O(n^2)$.

Answer 2:

We select $n_0 = 4$ and $c = \frac{1}{2}, \forall n \geq n_0$.

$$n^3 - 7n + 1 \geq \frac{1}{2}n^3 \text{ as } n \geq 4$$
$$= cn^3$$

Thus, by definition of big Ω, $n^3 - 7n + 1$ is in $\Omega(n^3)$.

Answer 3:

We select $n_0 = 1$ and $c = 6, \forall n \geq n_0$.

$$3n + 2 \leq 6n \text{ as } n \geq 1$$
$$= cn$$

Thus, by definition of big O, $3n + 2$ is in $O(n)$.
We select $n_0 = 1$ and $c = 1, \forall n \geq n_0$.

$$3n + 2 \geq n \text{ as } n \geq 1$$
$$= cn$$

Thus, by definition of big Ω, $3n + 2$ is in $\Omega(n)$. Therefore, by definition of big Θ, $3n + 2$ is in $\Theta(n)$.

Answer 4:

$f(n) = 7n + 8$ and $g(n) = n$. We select $n_0 = 1$ and $c = 15, \forall n \geq n_0$.

$$7n + 8 \leq 15n \text{ as } n \geq 1$$
$$= cn$$

Thus, by definition of big O, $7n + 8$ is in $O(n)$.

We select $n_0 = 1$ and $c = 7$, $\forall n \geq n_0$.

$7n + 8 \geq 7n$ as $n \geq 1$

$\qquad = cn$

Thus, by definition of big Ω, $7n + 8$ is in $\Omega(n)$. Therefore, by definition of big Θ, $7n + 8$ is in $\Theta(n)$.

Answer 5: The steps to analyze an algorithm in terms of asymptotic notations are described in the following. (i) Find out what the input is and what the input size n of the algorithm is? (ii) determine the maximum number of operations of the algorithm in terms of n, (iii) exclude all terms, except the highest order terms, and (iv) ignore all the constant factors.

Answer 6: An example of a problem in NP but not NP-Complete is the sorting of n numbers either in ascending or descending order. We can verify whether the list is sorted in polynomial time, and thus the problem is in NP. There are known algorithms to sort a list of numbers in polynomial time; the computational time complexity of merge sort is $O(n \log n)$. Thus the sorting problem is in P.

Answer 7: A problem is in NP class if (i) it is a decision problem and (ii) given an instance of the problem, it is verified whether the instance is a feasible solution in polynomial time.

Answer 8: Please check section 12.2.1.

Answer 9: We define the static spectrum reallocation for limited network operations (SSR-LNO) decision problem as:

Definition 13.1 Given a set of lightpaths $p \in P$, which can be in the primary or backup state and are initially allocated in the spectrum $F = \{0, \cdots, |F| - 1\}$, is it possible to reallocate all lightpaths with the highest used index at most k?

Theorem 13.1
The SSR-LNO decision problem is NP-complete [212].

Proof 13.1 The SSR-LNO decision problem is NP, as we can verify whether an instance of the SSR-LNO decision problem has at most k as the highest used index in polynomial time $O(1)$.

We show that the static lightpath establishment (SLE) problem, which is a known NP-complete problem [264], is reducible to the SSR-LNO decision problem. The SLE is defined as: is it possible to allocate a given set of lightpaths p with the highest used index at most h in a network $G(V, E)$?

First we construct an instance of the SSR-LNO decision problem from any instance of the SLE problem. An instance of the SLE problem consists of the set of lightpaths $p \in P$, the allocation index of the lightpaths f_p and the threshold h for the highest used index. An instance of the SSR-LNO decision problem is constructed with the following algorithm.

1. Define network $G'(V', E')$ with $|V'| = |V|$. Nodes $n', m' \in V'$ are connected by an edge in $G'(V', E')$ if and only if corresponding nodes $n, m \in V$ are connected by an edge in $G(V, E)$.

2. Define P' as the set of lightpaths in $G'(V', E')$. For each lightpath $p \in P$, define a corresponding lightpath $q \in P'$ that follows the same routing path in $G'(V', E')$ as p in $G(V, E)$. We set the state of all lightpaths $q \in P'$ as backup paths, which are reallocated without path exchanging.

3. Define the set of spectrum indexes $F' = \{0, \cdots, |F'| - 1\}$ with $|F'| = 2 \times |P|$.

4. For each lightpath $q \in P'$:

i. Assign an initial allocation index $f_q'^{\text{init}}$ above h. The spectrum range of F' is large enough to allocate all lightpaths above h. In the worse case scenario, where each spectrum index is used by a single lightpath, the required number of spectrum indexes is $|P| + h$ which is less than $|F'| = 2 \times |P|$, if $|P| > h$. The case with $|P| \leq h$ is trivial.

ii. Assign a reallocation index $f_q'^{\text{final}} = f_p$, with $p \in P$ the lightpath corresponding to $q \in P'$, as defined in (i).

5. Set $k = h$.

The described algorithm has a polynomial complexity of $O(|P|)$. It transforms any SLE instance into an SSR-LNO instance. The input of the defined SSR-LNO instance is the initial lightpath allocation in the spectrum F' and the output is the final lightpath allocation in the spectrum F'. Since all lightpaths are initialized as backup paths allocated with indexes above h, the transition from the initial state to the final state is performed in one transition step.

Consider that the SLE instance is a Yes instance. The highest used index is less or equal to h. By using the above described algorithm to define an SSR-LNO instance from any SLE instance, the highest used index by reallocated lightpaths in the SSR-LNO instance is the same as the highest used index by the lightpaths in the corresponding SLE instance. To each lightpath $q \in P'$, the algorithm assigns a reallocation index $f_q'^{\text{final}} = f_p$, where f_p is the allocation index of the corresponding lightpath $p \in P$. With corresponding lightpaths $p \in P$ and $q \in P'$ having the same routing path, respectively in $G(V, E)$ and $G'(V', E')$, the spectrum layout after reallocation in the SSR-LNO instance is the same as the spectrum layout in the SLE instance. Therefore the highest used index by the reallocated lightpaths in the SSR-LNO instance is less than or equal to $k = h$, which means that the SSR-LNO instance is a Yes instance.

Conversely, if the SSR-LNO instance is a Yes instance then the SLE instance is a Yes instance. Since k is defined to be equal to h as discussed above, the highest used index by the reallocated lightpaths in the SSR-LNO instance is the same as the highest used index by the lightpaths in the SLE instance.

We have confirmed that using the described transformation, if the SLE instance is a Yes instance, then the corresponding SSR-LNO instance is a Yes instance, and conversely. This proves that the SLE decision problem, a known NP-complete problem, is polynomial time reducible to the SSR-LNO decision problem. Thus, the SSR-LNO decision problem is NP-complete.

Note that the SSR-LNO instance defined by the transformation algorithm is made of only backup paths. For an instance with primary and backup paths in the 1+1 protection, we consider network $G''(V'', E'')$, which is an extension of network $G'(V', E')$, where $V'' = V'$, with an edge (s, d) added to connect source $s \in V'$ and destination $d \in V'$, for all source/destination pairs. The backup paths $q \in P'$ are defined in the same way as the transformation algorithm. For each lightpath $q \in P'$, its corresponding primary path is routed through the added edge between its source s and destination d, and is allocated using first-fit allocation. Edge $(s, d) \in E''$ is used only by the primary paths from source s to destination d. Since the first-fit allocation is used, edge $(s, d) \in E''$ does not present any fragmentation. As a result, the primary lightpaths are not touched during the defragmentation, and the highest spectrum index used by a primary lightpath is at most h. If the backup lightpaths can be reallocated with the highest used index being at most h, then the SSR-LNO instance is a Yes instance.

Answer 10: Logarithmic algorithm — $O(\log n)$ — binary search. Linear algorithm — $O(n)$ — linear search. Superlinear algorithm — $O(n \log n)$ — heap sort and merge sort. Polynomial algorithm — $O(n^c)$ — strassen's matrix multiplication, bubble sort, selection sort, insertion sort, and bucket sort. Exponential algorithm — $O(c^n)$ — tower of Hanoi. Factorial algorithm — $O(n!)$ — determinant expansion by minors and brute force search algorithm for traveling salesman problem.

Chapter 13

Answer 1: To achieve a long transmission reach, optical signals must be amplified at periodic regeneration points along the fiber span to compensate for the power loss experienced in fibers. The amplification of signal is typically performed quickly to avoid delay by transient-suppressed erbium doped fiber amplifiers (TS-EDFAs). At the current time, TS-EDFAs are available for single core fibers. On the other hand, TS-EDFAs are unavailable for multi-core fibers. For amplification of multi-core fiber signals, one technique is to demultiplex the multi-core fiber signals into multiple single-core fibers and then amplify the signals in each fiber using conventional single-core TS-EDFAs. The amplified signals are then recombined and injected back into the multi-core fiber span, which make the system complicated. Therefore, amplification of multi-core fiber signals is more challenging than that of single-core fiber signals.

Answer 2: Hop retuning technology is executed with the help of high-performance-sophisticated devices and components, including rapidly-tunable lasers and athermal arrayed waveguide grating (AWG). In finely-granular systems, such as 2.5 GHz systems, due to the lack of adequate filtering components, the hop retuning approach is unsuitable.

Answer 3: The EON allocates spectral resources with just enough bandwidth to satisfy the traffic demands. However, traffic grooming is still essential in EONs for the following reasons, (i) BVT is normally designed so as to maximize the traffic rate in the network, and it does not support slicing at a very early stage. Electrical traffic grooming is applied in order to use transponder capacity efficiently. (ii) Generally speaking, a filter guard band between two adjacent channels should be assigned to resolve optical filter issues. Traffic grooming can minimize filter guard band usage by aggregating traffic electrically. The electrical switching fabric is still needed for traffic grooming in the EON, similar to WDM networks.

Answer 4: Please check section 13.5.

Answer 5: The effect of physical layer impairments increases with increase in data rate. EONs can support a higher data rate than WDM based optical networks. Therefore, managing physical layer impairments in EONs is more challenging than that of WDM based optical networks.

References

[1] K. I. Sato, S. Okamoto, and H. Hadama. Network performance and integrity enhancement with optical path layer technologies. *IEEE Journal on Selected Areas in Communications*, 12(1):159–170, 1994.

[2] S. A. Calta, E. Loizides, R. N. Strangwayes, and J. A. deVeer. Enterprise systems connection (ESCON) architecture-system overview. *IBM Journal of Research and Development*, 36(4):535–551, 1992.

[3] F. E. Ross. An overview of FDDI: The fiber distributed data interface. *IEEE Journal on Selected Areas in Communications*, 7(7):1043–1051, 1989.

[4] W. Bux, F. Closs, K. Kuemmerle, H. Keller, and H. Mueller. Architecture and design of a reliable token-ring network. *IEEE Journal on Selected Areas in Communications*, 1(5):756–765, 1983.

[5] R. M. Metcalfe and D. R. Boggs. Ethernet: Distributed packet switching for local computer networks. *Communications of the ACM*, 19(7):395–404, 1976.

[6] B. Mukherjee. *Optical WDM Networks*. Springer, 2006.

[7] C. S. R. Murthy and G. Mohan. *WDM Optical Networks: Concepts, Design and Algorithms*. PHI, 2003.

[8] C. V. Saradhi and S. Subramaniam. Physical layer impairment aware routing (PLIAR) in WDM optical networks: Issues and challenges. *Communications Surveys & Tutorials, IEEE*, 11(4):109–130, 2009.

[9] G Keiser. *Optical Fiber Communications*. McGraw-Hill, 1991.

[10] J. Strand, A. Chiu, and R. Tkach. Issues for routing in the optical layer. *IEEE Communication Magazine*, 39(2):81–87, 2001.

[11] G. P. Agrawal and N. A. Olsson. Self-phase modulation and spectral broadening of optical pulses in semiconductor laser amplifiers. *IEEE Journal of Quantum Electronics*, 25(11):2297–2306, 1989.

[12] N. Imoto, H. A. Haus, and Y. Yamamoto. Quantum nondemolition measurement of the photon number via the optical kerr effect. *Physical Review A*, 32(4):2287, 1985.

[13] R. K. Jain and R. C. Lind. Degenerate four-wave mixing in semiconductor-doped glasses. *JOSA*, 73(5):647–653, 1983.

[14] G. P. Agrawal. Modulation instability induced by cross-phase modulation. *Physical Review Letters*, 59(8):880, 1987.

[15] E. P. Ippen and R. H. Stolen. Stimulated brillouin scattering in optical fibers. *Applied Physics Letters*, 21(11):539–541, 1972.

[16] K. J. Blow and D. Wood. Theoretical description of transient stimulated raman scattering in optical fibers. *IEEE Journal of Quantum Electronics*, 25(12):2665–2673, 1989.

[17] R. Ramaswami, K. N. Sivarajan, and G. H. Sasaki. *Optical Networks: A Practical Perspective*. Morgan Kaufmann, 2009.

[18] N. S. Kapov. *Heuristic algorithms for virtual topology design and routing and wavelength assignment in WDM networks*. PhD thesis, PhD in Philosophy, University of Zagreb, Zagreb, Croatia, 2006.

[19] A. S. Tanenbaum. *Computer Networks*. Prentice-Hall Englewood Cliffs, 1989.

[20] B. C. Chatterjee, N. Sarma, and P. P. Sahu. Priority based routing and wavelength assignment with traffic grooming for optical networks. *Journal of Optical Communications and Networking*, 4(6):480–489, 2012.

[21] B. C. Chatterjee, N. Sarma, and P. P. Sahu. Priority based dispersion-reduced wavelength assignment for optical networks. *Journal of Lightwave Technology*, 31(2):257–263, 2013.

[22] K. Zhu, H. Zang, and B. Mukherjee. A comprehensive study on next-generation optical grooming switches. *Selected Areas in Communications, IEEE Journal on*, 21(7):1173–1186, 2003.

[23] B. C. Chatterjee, N. Sarma, and P. P. Sahu. A heuristic priority based wavelength assignment scheme for optical networks. *Optik-International Journal for Light and Electron Optics*, 123(17):1505–1510, 2012.

[24] B. C. Chatterjee, N. Sarma, P. P. Sahu, and E. Oki. *Routing and Wavelength Assignment for WDM-based Optical Networks: Quality-of-Service and Fault Resilience*, Volume 410. Springer, 2016.

[25] How much bandwidth do we need? http://arstechnica.com/business/2012/05/bandwidth-explosion-as-internet -use-soars-can-bottlenecks-be-averted/, July 2014.

[26] M. Jinno, H. Takara, and B. Kozicki. Dynamic optical mesh networks: Drivers, challenges and solutions for the future. In *Optical Communication, 2009. ECOC'09. 35th European Conference on*, pages 1–4. IEEE, 2009.

[27] S. Roy, A. Malik, A. Deore, S. Ahuja, O. Turkcu, S. Hand, and S. Melle. Evaluating efficiency of multi-layer switching in future optical transport networks. In *National Fiber Optic Engineers Conference*. Optical Society of America, 2013.

[28] M. Jinno, H. Takara, B. Kozicki, Y. Tsukishima, Y. Sone, and S. Matsuoka. Spectrum-efficient and scalable elastic optical path network: Architecture, benefits, and enabling technologies. *Communications Magazine, IEEE*, 47(11):66–73, 2009.

[29] M. Jinno, B. Kozicki, H. Takara, A. Watanabe, Y. Sone, T. Tanaka, and A. Hirano. Distance-adaptive spectrum resource allocation in spectrum-sliced elastic optical path network [Topics in Optical Communications]. *Communications Magazine, IEEE*, 48(8):138–145, 2010.

[30] O. Gerstel, M. Jinno, A. Lord, and S. J. B. Yoo. Elastic optical networking: A new dawn for the optical layer? *Communications Magazine, IEEE*, 50(2):s12–s20, 2012.

[31] K. Christodoulopoulos, I. Tomkos, and E. A. Varvarigos. Elastic bandwidth allocation in flexible OFDM-based optical networks. *Journal of Lightwave Technology*, 29(9):1354–1366, 2011.

[32] G. Zhang, M. D. Leenheer, A. Morea, and B. Mukherjee. A survey on OFDM-based elastic core optical networking. *IEEE Communication Surveys & Tutorials*, 15(1):65–87, 2013.

[33] E. Oki and B. C. Chatterjee. Design and control in elastic optical networks: issues, challenges, and research directions. In *proc. IEEE ICNC*, pages 1–5. IEEE, 2017.

[34] M. Jinno. Elastic optical networking: Roles and benefits in beyond 100-gb/s era. *Journal of Lightwave Technology*, 35(5):1116–1124, 2016.

[35] M. Jinno, H. Takara, B. Kozicki, Y. Tsukishima, T. Yoshimatsu, T. Kobayashi, Y. Miyamoto, K. Yonenaga, A. Takada, O. Ishida and S. Matsuoka. Demonstration of novel spectrum-efficient elastic optical path network with per-channel variable capacity of 40 Gb/s to over 400 Gb/s. In *Optical Communication, 2008. ECOC 2008. 34th European Conference on*, pages 1–2. IEEE, 2008.

[36] J. Armstrong. OFDM for optical communications. *Journal of Lightwave Technology*, 27(3):189–204, 2009.

[37] R. V. Nee and R. Prasad. *OFDM for Wireless Multimedia Communications*. Artech House, Inc., 2000.

[38] Q. Yang, W. Shieh, and Y. Ma. Bit and power loading for coherent optical OFDM. In *OECC/ACOFT 2008-Joint Conference of the Opto-Electronics and Communications Conference and the Australian Conference on Optical Fibre Technology*, pages 1–2. IEEE, 2008.

[39] H. Takara, B. Kozicki, Y. Sone, T. Tanaka, A. Watanabe, A. Hirano, K. Yonenaga, and M. Jinno. Distance-adaptive super-wavelength routing in elastic optical path network (SLICE) with optical OFDM. In *36th European Conference and Exhibition on Optical Communication*, pages 1–3. IEEE, 2010.

[40] G.-H. Gho, L. Klak, and J. M. Kahn. Rate-adaptive coding for optical fiber transmission systems. *Journal of Lightwave Technology*, 29(2):222–233, 2011.

[41] O. Rival, A. Morea, and J. C. Antona. Optical network planning with rate-tunable NRZ transponders. In *2009 35th European Conference on Optical Communication*, pages 1–2. IEEE, 2009.

[42] B. Kozicki, H. Takara, T. Yoshimatsu, K. Yonenaga, and M. Jinno. Filtering characteristics of highly-spectrum efficient spectrum-sliced elastic optical path (SLICE) network. In *National Fiber Optic Engineers Conference*, page JWA43. Optical Society of America, 2009.

[43] V. López and L. Velasco (eds.). Elastic optical networks. *Architectures, Technologies, and Control,* Switzerland: Springer Int. Publishing, 2016.

[44] B. Kozicki, H. Takara, Y. Tsukishima, T. Yoshimatsu, K. Yonenaga, and M. Jinno. Experimental demonstration of spectrum-sliced elastic optical path network (slice). *Optics Express*, 18(21):22105–22118, 2010.

[45] M. Jinno, Y. Miyamoto, and Y. Hibino. Networks: optical-transport networks in 2015. *Nature Photonics*, 1(3):157, 2007.

[46] M. Jinno, H. Takara, Y. Sone, K. Yonenaga, and A. Hirano. Multiflow optical transponder for efficient multilayer optical networking. *Communications Magazine, IEEE*, 50(5):56–65, 2012.

[47] A. Lord, P. Wright, and A. Mitra. Core networks in the flexgrid era. *Journal of Lightwave Technology*, 33(5):1126–1135, 2015.

[48] G. Bosco, A. Carena, V. Curri, P. Poggiolini, and F. Forghieri. Performance limits of Nyquist-WDM and Co-OFDM in high-speed PM-QPSK systems. *IEEE Photonics Technology Letters*, 22(15):1129–1131, 2010.

[49] G. Bosco, V. Curri, A. Carena, P. Poggiolini, and F. Forghieri. On the performance of nyquist-wdm terabit superchannels based on PM-BPSK, PM-QPSK, PM-8QAM or PM-16QAM subcarriers. *Journal of Lightwave Technology*, 29(1):53–61, 2011.

[50] T. Tanaka, A. Hirano, and M. Jinno. Impact of transponder architecture on the scalability of optical nodes in elastic optical networks. *IEEE Communications Letters*, 17(9):1846–1848, 2013.

[51] N. Sambo, P. Castoldi, A. D'Errico, E. Riccardi, A. Pagano, M. S. Moreolo, J. M. Fabrega, D. Rafique, A. Napoli, S. Frigerio, E. H. Salas, G. Zervas, M. Nolle, J. K. Fischer, A. Lord and J. P. F. P. Gimenez. Next generation sliceable bandwidth variable transponders. *IEEE Communications Magazine*, 53(2):163–171, 2015.

[52] A. Napoli, P. W. Berenguer, T. Rahman, G. Khanna, M. M. Mezghanni, L. Gardian, E. Riccardi, A. C. Piat, S. Calabrò, S. Dris, A. Richter, J. K. Fischer, B. Sommerkorn-Krombholz and B. Spinnler. Digital pre-compensation techniques enabling high-capacity bandwidth variable transponders. *Optics Communications*, 409:52–65, 2018.

[53] N. G. Montoro, J. M. Finochietto, and A. Bianco. Translucent provisioning in elastic optical networks with sliceable bandwidth variable transponders. In *2018 IEEE Global Communications Conference (GLOBECOM)*, pages 1–6. IEEE, 2018.

[54] M. Zhu, Q. Sun, S. Zhang, P. Gao, B. Chen, and J. Gu. Energy-aware virtual optical network embedding in sliceable-transponder-enabled elastic optical networks. *IEEE Access*, 7:41897–41912, 2019.

[55] L. M. González, S. V. Heide, X. Xue, J. V. Weerdenburg, N. Calabretta, C. Okonkwo, J. Fàbrega, and M. S. Moreolo. Programmable adaptive BVT for future optical metro networks adopting SOA-based switching nodes. In *Photonics*, volume 5, page 24. Multidisciplinary Digital Publishing Institute, 2018.

[56] N. Sambo, A. DErrico, C. Porzi, V. Vercesi, M. Imran, F. Cugini, A. Bogoni, L. Potì, and P. Castoldi. Sliceable transponder architecture including multiwavelength source. *Journal of Optical Communications and Networking*, 6(7):590–600, 2014.

[57] V. López, B. D. Cruz, Ó. G. D. Dios, O. Gerstel, N. Amaya, G. Zervas, D. Simeonidou, and J. P. F. Palacios. Finding the target cost for sliceable bandwidth variable transponders. *Journal of Optical Communications and Networking*, 6(5):476–485, 2014.

[58] J. Zhang, Y. Ji, M. Song, Y. Zhao, X. Yu, and B. Mukherjee. Dynamic traffic grooming in sliceable bandwidth-variable transponder enabled elastic optical networks. *Journal of Lightwave Technology (accepted)*, 2015.

[59] V. Lopez, O. G. D. Dios, O. Gerstel, N. Amaya, G. Zervas, D. Simeonidou, and J. P. F. Palacios. Target cost for sliceable bandwidth variable transponders in a real core network. In *Future Network and Mobile Summit (FutureNetworkSummit), 2013*, pages 1–9. IEEE, 2013.

[60] G. Baxter, S. Frisken, D. Abakoumov, H. Zhou, I. Clarke, A. Bartos, and S. Poole. Highly programmable wavelength selective switch based on liquid crystal on silicon switching elements. In *Optical Fiber Communication Conference*, page OTuF2. Optical Society of America, 2006.

[61] W. Kabaciński, A. A. Tameemie, and R. Rajewski. Rearrangeability of wavelength-space-wavelength switching fabric architecture for elastic optical switches. *IEEE Access*, 7:64993–65006, 2019.

[62] S. Frisken, G. Baxter, D. Abakoumov, H. Zhou, I. Clarke, and S. Poole. Flexible and grid-less wavelength selective switch using LCOS technology. In *Optical Fiber Communication Conference*, page OTuM3. Optical Society of America, 2011.

[63] R. Ryf, Y. Su, L. Moller, S. Chandrasekhar, X. Liu, D. T. Neilson, and C. R. Giles. Wavelength blocking filter with flexible data rates and channel spacing. *Lightwave Technology, Journal of*, 23(1):54–61, 2005.

[64] O. Rival and A. Morea. Elastic optical networks with 25–100G format-versatile WDM transmission systems. In *OptoeElectronics and Communications Conference (OECC), 2010 15th*, pages 100–101. IEEE, 2010.

[65] N. Amaya, G. Zervas, and D. Simeonidou. Introducing node architecture flexibility for elastic optical networks. *Journal of Optical Communications and Networking*, 5(6):593–608, 2013.

[66] W. Kabaciński, M. Abdulsahib, and M. Michalski. Wide-sense nonblockingw-sw node architectures for elastic optical networks. *IEICE Transactions on Communications*, 2018.

[67] A. Kadohata, A. Hirano, M. Fukutoku, T. Ohara, Y. Sone, and O. Ishida. Multi-layer greenfield re-grooming with wavelength defragmentation. *IEEE Commun. Lett.*, 16(4):530–532, 2012.

[68] M. Zhang, W. Shi, L. Gong, W. Lu, and Z. Zhu. Bandwidth defragmentation in dynamic elastic optical networks with minimum traffic disruptions. In *IEEE International Conference on Communications (ICC)*, 3894–3898. IEEE, 2013.

[69] M. Zhang, C. You, H. Jiang, and Z. Zhu. Dynamic and adaptive bandwidth defragmentation in spectrum-sliced elastic optical networks with time-varying traffic. *J. of Lightwave Technol.*, 32(5):1014–1023, 2014.

[70] N. Amaya, G. Zervas, and D. Simeonidou. Architecture on demand for transparent optical networks. In *Transparent Optical Networks (ICTON), 2011 13th International Conference on*, 1–4. IEEE, 2011.

[71] M. Klinkowski and K. Walkowiak. Routing and spectrum assignment in spectrum sliced elastic optical path network. *IEEE Communications Letters*, 15(8):884–886, 2011.

[72] B. C. Chatterjee, N. Sarma, and E. Oki. Routing and spectrum allocation in elastic optical networks: A tutorial. *IEEE Communications Surveys & Tutorials*, 17(3):1776–1800, 2015.

[73] S. Talebi, F. Alam, I. Katib, M. Khamis, R. Salama, and G. N. Rouskas. Spectrum management techniques for elastic optical networks: A survey. *Optical Switching and Networking*, 13:34–48, 2014.

[74] J. Wu, S. Subramaniam, and H. Hasegawa. Efficient dynamic routing and spectrum assignment for multifiber elastic optical networks. *IEEE/OSA Journal of Optical Communications and Networking*, 11(5):190–201, 2019.

[75] S. Behera, A. Deb, G. Das, and B. Mukherjee. Impairment aware routing, bit loading, and spectrum allocation in elastic optical networks. *Journal of Lightwave Technology*, 37(13):3009–3020, 2019.

[76] M. Salani, C. Rottondi, and M. Tornatore. Routing and spectrum assignment integrating machine-learning-based QoT estimation in elastic optical networks. In *IEEE INFOCOM 2019-IEEE Conference on Computer Communications*, 1738–1746. IEEE, 2019.

[77] B. C. Chatterjee, N. Sarma, and P. P. Sahu. Review and performance analysis on routing and wavelength assignment approaches for optical networks. *IETE Technical Review*, 30(1):12–23, 2013.

[78] Y. Wang, X. Cao, and Y. Pan. A Study of the routing and spectrum allocation in spectrum-sliced elastic optical path networks. In *INFOCOM, 2011
Proceedings IEEE*, 1503–1511. IEEE, 2011.

[79] S. Subramaniam and R.A. Barry. Wavelength assignment in fixed rout-
ing WDM networks. In *the Proceedings of International Conference on
Communications (ICC-97)*, 406–410, Montreal, Canada, 12-12 June 1997.
IEEE.

[80] R. Ramamurthy and B. Mukherjee. Fixed-alternate routing and wavelength conversion in wavelength-routed optical networks. *Networking,
IEEE/ACM Transactions on Networking*, 10(3):351–367, 2002.

[81] J.P. Jue and G. Xiao. An adaptive routing algorithm for wavelength-routed
optical networks with a distributed control scheme. In *the Proceedings of
Ninth International Conference on Computer Communications and Networks*, 192–197. IEEE, 2002.

[82] A. Castro, L. Velasco, M. Ruiz, M.A. Klinkowski, J.P. FernáNdez-
Palacios, and D. Careglio. Dynamic routing and spectrum (re)allocation in
future flexgrid optical networks. *Computer Networks*, 56(12):2869–2883,
2012.

[83] X. Wan, N. Hua, and X. Zheng. Dynamic routing and spectrum assignment in spectrum-flexible transparent optical networks. *Optical Communications and Networking, IEEE/OSA Journal of*, 4(8):603–613, 2012.

[84] K. Chan and T.S.P. Yum. Analysis of least congested path routing in
WDM lightwave networks. In *the Proceedings of The 13th International Conference on Computer Communications (INFOCOM-94)*, 962–
969. IEEE, 1994.

[85] T. H. Cormen. *Introductions to Algorithms*. McGraw-Hill Companies,
2003.

[86] B. C. Chatterjee, N. Sarma, and P. P. Sahu. A priority based wavelength
assignment scheme for optical network. In *International workshop on
Network Modeling and Analysis (IWNMA-2011)*, 1–6, Bangalore, India,
2011.

[87] B. C. Chatterjee, N. Sarma, and P. P. Sahu. Dispersion-reduction routing
and wavelength assignment for optical networks. In *proceeding of the 2nd
International Conference on Trends in Optics and Photonics*, 456–463,
2011.

[88] X. Chen, A. Jukan, and A. Gumaste. Multipath de-fragmentation: Achieving better spectral efficiency in elastic optical path networks. In *INFO-
COM, 2013 Proceedings IEEE*, 390–394. IEEE, 2013.

[89] L. Ruan and N. Xiao. Survivable multipath routing and spectrum allocation in OFDM-based flexible optical networks. *Journal of Optical Communications and Networking*, 5(3):172–182, 2013.

[90] L. Ruan and Y. Zheng. Dynamic survivable multipath routing and spectrum allocation in OFDM-based flexible optical networks. *Journal of Optical Communications and Networking*, 6(1):77–85, 2014.

[91] F. Yousefi and A. G. Rahbar. Novel fragmentation-aware algorithms for multipath routing and spectrum assignment in elastic optical networks-space division multiplexing (EON-SDM). *Optical Fiber Technology*, 46:287–296, 2018.

[92] M. N. Dharmaweera, J. Zhao, L. Yan, M. Karlsson, and E. Agrell. Traffic-grooming-and multipath-routing-enabled impairment-aware elastic optical networks. *Journal of Optical Communications and Networking*, 8(2):58–70, 2016.

[93] Z. Fan, Y. Qiu, and C. K. Chan. Dynamic multipath routing with traffic grooming in OFDM-based elastic optical path networks. *Journal of Lightwave Technology*, 33(1):275–281, 2015.

[94] M. Klinkowski, M. Ruiz, L. Velasco, D. Careglio, V. Lopez, and J. Comellas. Elastic spectrum allocation for time-varying traffic in flexgrid optical networks. *Selected Areas in Communications, IEEE Journal on*, 31(1):26–38, 2013.

[95] A. A. Garcia. Elastic spectrum allocation in flexgrid optical networks. Technical report, Universitat Politcnica de Catalunya, 2012.

[96] L. Berger. Generalized multi-protocol label switching (GMPLS) signaling resource reservation protocol-traffic engineering (RSVP-TE) extensions. http://tools.ietf.org/html/rfc3473, July 2014.

[97] B. C. Chatterjee and E. Oki. Performance evaluation of spectrum allocation policies for elastic optical networks. In *2015 17th International Conference on Transparent Optical Networks (ICTON)*, 1–4. IEEE, 2015.

[98] A. Rosa, C. Cavdar, S. Carvalho, J. Costa, and L. Wosinska. Spectrum Allocation policy modeling for elastic optical networks. In *High Capacity Optical Networks and Enabling Technologies (HONET), 2012 9th International Conference on*, 242–246. IEEE, 2012.

[99] R. Wang and B. Mukherjee. Spectrum management in heterogeneous bandwidth optical networks. *Optical Switching and Networking*, 11:83–91, 2014.

[100] W. Fadini and E. Oki. A subcarrier-slot partition scheme for wavelength assignment in elastic optical networks. In *IEEE International Conference on High Performance Switching and Routing (HPSR)*, 7–12. IEEE, 2014.

[101] X. Liu, L. Gong, and Z. Zhu. Design integrated RSA for multicast in elastic optical networks with a layered approach. In *Global Communications Conference (GLOBECOM), 2013 IEEE*, 2346–2351. IEEE, 2013.

[102] Y. Yin, H. Zhang, M. Zhang, M. Xia, Z. Zhu, S. Dahlfort, and S. J. B. Yoo. Spectral and spatial 2D fragmentation-aware routing and spectrum assignment algorithms in elastic optical networks [Invited]. *Journal of Optical Communications and Networking*, 5(10):A100–A106, 2013.

[103] M. Zhang, W. Shi, L. Gong, W. Lu, and Z. Zhu. Bandwidth defragmentation in dynamic elastic optical networks with minimum traffic disruptions. In *Communications (ICC), 2013 IEEE International Conference on*, 3894–3898. IEEE, 2013.

[104] X. Zhou, W. Lu, L. Gong, and Z. Zhu. Dynamic RMSA in elastic optical networks with an adaptive genetic algorithm. In *Global Communications Conference (GLOBECOM), 2012 IEEE*, 2912–2917. IEEE, 2012.

[105] Y. Yin, Z. Zhu, S. J. Yoo, and M. Zhang. Fragmentation-aware routing, modulation and spectrum assignment algorithms in elastic optical networks. In *Optical Fiber Communication Conference*, OW3A–5. Optical Society of America, 2013.

[106] J. Zhao, B. Bao, H. Yang, E. Oki, and B. C. Chatterjee. Holding-time- and impairment-aware shared spectrum allocation in mixed-line-rate elastic optical networks. *IEEE/OSA Journal of Optical Communications and Networking*, 11(6):322–332, 2019.

[107] E. E. Moghaddam, H. Beyranvand, and J. A. Salehi. Routing, spectrum and modulation level assignment, and scheduling in survivable elastic optical networks supporting multi-class traffic. *Journal of Lightwave Technology*, 36(23):5451–5461, 2018.

[108] J. Zhao, B. Bao, B. C. Chatterjee, E. Oki, J. Hu, and D. Ren. Dispersion based highest-modulation-first last-fit spectrum allocation scheme for elastic optical networks. *IEEE Access*, 6:59907–59916, 2018.

[109] B. C. Chatterjee and E. Oki. Dispersion-adaptive first-last fit spectrum allocation scheme for elastic optical networks. *IEEE Communications Letters*, 20(4):696–699, 2016.

[110] B. C. Chatterjee and E. Oki. Lightpath threshold adaptation algorithm for dispersion-adaptive first-last fit spectrum allocation scheme in elastic

optical networks. In *2016 18th International Conference on Transparent Optical Networks (ICTON)*, 1–4. IEEE, 2016.

[111] P. S. Khodashenas, J. Comellas, S. Spadaro, J. Perelló, and G. Junyent. Using spectrum fragmentation to better allocate time-varying connections in elastic optical networks. *Journal of Optical Communications and Networking*, 6(5):433–440, 2014.

[112] Y. Sone, A. Hirano, A. Kadohata, M. Jinno, and O. Ishida. Routing and spectrum assignment algorithm maximizes spectrum utilization in optical networks. In *European Conference and Exposition on Optical Communications*, 1–3. Optical Society of America, 2011.

[113] A. N. Patel, P. N. Ji, J. P. Jue, and T. Wang. Routing, wavelength assignment, and spectrum allocation algorithms in transparent flexible optical WDM networks. *Optical Switching and Networking*, 9(3):191–204, 2012.

[114] R. Wang and B. Mukherjee. Spectrum management in heterogeneous bandwidth networks. *Proc. IEEE Globecom*, 2012.

[115] Z. Ding, Z. Xu, X. Zeng, T. Ma, and F. Yang. Hybrid routing and spectrum assignment algorithms based on distance-adaptation combined co-evolution and heuristics in elastic optical networks. *Optical Engineering*, 53(4):046105–046105, 2014.

[116] T. Takagi, H. Hasegawa, K. Sato, T. Tanaka, B. Kozicki, Y. Sone, and M. Jinno. Algorithms for maximizing spectrum efficiency in elastic optical path networks that adopt distance adaptive modulation. In *Proc. ECOC*, 2010.

[117] T. Takagi, H. Hasegawa, K. I. Sato, Y. Sone, B. Kozicki, A. Hirano, and M. Jinno. Dynamic routing and frequency slot assignment for elastic optical path networks that adopt distance adaptive modulation. In *Optical Fiber Communication Conference*, OTuI7. Optical Society of America, 2011.

[118] S. Yang and F. Kuipers. Impairment-aware routing in translucent spectrum-sliced elastic optical path networks. In *Networks and Optical Communications (NOC), 2012 17th European Conference on*, 1–6. IEEE, 2012.

[119] H. Beyranvand and J. A. Salehi. A quality-of-transmission aware dynamic routing and spectrum assignment scheme for future elastic optical networks. *Journal of Lightwave Technology*, 31(18):3043–3054, 2013.

[120] A.L. Chiu and E.H. Modiano. Traffic grooming algorithms for reducing electronic multiplexing costs in WDM ring networks. *Lightwave Technology, IEEE/OSA Journal of*, 18(1):2–12, 2000.

[121] J. Wang, W. Cho, V.R. Vemuri, and B. Mukherjee. Improved approaches for cost-effective traffic grooming in WDM ring networks: ILP formulations and single-hop and multi-hop connections. *Journal of Lightwave Technology, IEEE/OSA Journal of*, 19(11):1645–1653, 2001.

[122] M. Liu, M. Tornatore, and B. Mukherjee. Survivable traffic grooming in elastic optical networks — Shared path protection. In *the Proceedings of International Conference on Communications (ICC-12)*, 6230–6234. IEEE, 2012.

[123] S. Zhang, C. Martel, and B. Mukherjee. Dynamic traffic grooming in elastic optical networks. *Selected Areas in Communications, IEEE Journal on*, 31(1):4–12, 2013.

[124] K. Zhu and B. Mukherjee. A review of traffic grooming in WDM optical networks: Architectures and challenges. *Optical Networks Magazine*, 4(2):55–64, 2003.

[125] K. Zhu and B. Mukherjee. Traffic grooming in an optical WDM mesh network. *Selected Areas in Communications, IEEE Journal on*, 20(1):122–133, 2002.

[126] E. Modiano. Traffic grooming in WDM networks. *Communications Magazine, IEEE*, 39(7):124–129, 2001.

[127] Y. Zhang, X. Zheng, Q. Li, N. Hua, Y. Li, and H. Zhang. Traffic grooming in spectrum-elastic optical path networks. In *Proc. OFC/NFOEC*, 2011.

[128] G. Zhang, M. D. Leenheer, and B. Mukherjee. Optical traffic grooming in OFDM-based elastic optical networks [Invited]. *Journal of Optical Communications and Networking*, 4(11):B17–B25, 2012.

[129] S. Zhang, M. Tornatore, G. Shen, J. Zhang, and B. Mukherjee. Evolving traffic grooming in multi-layer flexible-grid optical networks with software-defined elasticity. *Lightwave Technology, Journal of*, 32(16):2905–2914, 2014.

[130] B. Kozicki, H. Takara, and M. Jinno. Enabling technologies for adaptive resource alocation in elastic optical path network (SLICE). In *Communications and Photonics Conference and Exhibition (ACP), 2010 Asia*, 23–24, 2010.

[131] B. J. Zhang, Y. Zhao, X. Yu, J. Zhang, B. Mukherjee. Auxiliary graph model for dynamic traffic grooming in elastic optical networks with sliceable optical transponder. In *Optical Communication (ECOC), 2014 European Conference on*, 1–3, 2014.

[132] P. Winzer. Beyond 100G ethernet. *Communications Magazine, IEEE,* 48(7):26–30, 2010.

[133] W. Fawaz and K. Chen. Survivability-oriented Quality of Service in Optical Networks. In A. Mellouk, editor, *Quality of Service Engineering in Next Generation Heterogenous Networks,* 197–211. Wiley Online Library, London, UK, 2010.

[134] E. Bouillet, G. Ellinas, J. F. Labourdette, and R. Ramamurthy. *Path Routing in Mesh Optical Networks.* Wiley Online Library, 2007.

[135] M. Liu, M. Tornatore, and B. Mukherjee. Survivable traffic grooming in elastic optical networks — Shared protection. *Journal of Lightwave Technology,* 31(6):903–909, 2013.

[136] G. Shen, Y. Wei, and S. K. Bose. Optimal design for shared backup path protected elastic optical networks under single-link failure. *Journal of Optical Communications and Networking,* 6(7):649–659, 2014.

[137] K. Walkowiak, M. Klinkowski, B. Rabiega, and R. Goścień. Routing and spectrum allocation algorithms for elastic optical networks with dedicated path protection. *Optical Switching and Networking,* 13:63–75, 2014.

[138] J. Wu, Y. Liu, C. Yu, and Y. Wu. Survivable routing and spectrum allocation algorithm based on P-cycle protection in elastic optical networks. *Optik-International Journal for Light and Electron Optics,* 2014.

[139] M. Klinkowski and K. Walkowiak. Offline RSA algorithms for elastic optical networks with dedicated path protection consideration. In *Ultra Modern Telecommunications and Control Systems and Workshops (ICUMT), 2012 4th International Congress on,* 670–676, Oct 2012.

[140] A. Giorgetti, F. Paolucci, F. Cugini, and P. Castoldi. Fast restoration in SDN-based flexible optical networks. In *Optical Fiber Communication Conference,* Th3B–2. Optical Society of America, 2014.

[141] Y. Wei, G. Shen, and S. K. Bose. Span-restorable elastic optical networks under different spectrum conversion capabilities. 2014.

[142] B. Chen, J. Zhang, Y. Zhao, C. Lv, w. Zhang, S. Huang, X. Zhang, and W. Gu. Multi-link failure restoration with dynamic load balancing in spectrum-elastic optical path networks. *Optical Fiber Technology,* 18(1):21–28, 2012.

[143] Y. Sone, A. Watanabe, W. Imajuku, Y. Tsukishima, B. Kozicki, H. Takara, and M. Jinno. Highly survivable restoration scheme employing optical

bandwidth squeezing in spectrum-sliced elastic optical path (SLICE) network. In *Optical Fiber Communication Conference*, page OThO2. Optical Society of America, 2009.

[144] Y. Sone, A. Watanabe, W. Imajuku, Y. Tsukishima, B. Kozicki, H. Takara, and M. Jinno. Bandwidth squeezed restoration in spectrum-sliced elastic optical path networks (SLICE). *Journal of Optical Communications and Networking*, 3(3):223–233, 2011.

[145] F. Paolucci, A. Castro, F. Cugini, L. Velasco, and P. Castoldi. Multipath restoration and bitrate squeezing in SDN-based elastic optical networks [Invited]. *Photonic Network Communications*, 1–13, 2014.

[146] F. Ji, X. Chen, W. Lu, J. J. P. C. Rodrigues, and Z. Zhu. Dynamic *P*-cycle configuration in spectrum-sliced elastic optical networks. In *Global Communications Conference (GLOBECOM), 2013 IEEE*, 2170–2175. IEEE, 2013.

[147] G. Shen and R. Tucker. Energy-minimized design for IP over WDM networks. *Optical Communications and Networking, IEEE/OSA Journal of*, 1(1):176–186, 2009.

[148] W. V. Heddeghem, F. Idzikowski, W. Vereecken, D. Colle, M. Pickavet, and P. Demeester. Power consumption modeling in optical multilayer networks. *Photonic Network Communications*, 24(2):86–102, 2012.

[149] J. Zhang, Y. Zhao, J. Zhang, and B. Mukherjee. Energy efficiency of IP-over-elastic optical networks with sliceable optical transponder. In *Optical Fiber Communication Conference*, pages W3A–4. Optical Society of America, 2014.

[150] J. Zhang, Y. Zhao, X. Yu, J. Zhang, M. Song, Y. Ji, and B. Mukherjee. Energy-efficient traffic grooming in sliceable-transponder-equipped IP-over-elastic optical networks [Invited]. *Journal of Optical Communications and Networking*, 7(1):A142–A152, 2015.

[151] A. Fallahpour, H. Beyranvand, S. A. Nezamalhosseini, and J. A. Salehi. Energy efficient routing and spectrum assignment with regenerator placement in elastic optical networks. *Journal of Lightwave Technology*, 32(10):2019–2027, 2014.

[152] S. Zhang and B. Mukherjee. Energy-efficient dynamic provisioning for spectrum elastic optical networks. In *Communications (ICC), 2012 IEEE International Conference on*, 3031–3035. IEEE, 2012.

[153] M. Chino, T. Miyazaki, E. Oki, S. Okamoto, and N. Yamanaka. Adaptive elastic spectrum allocation based on traffic fluctuation estimate in flexible OFDM-based optical networks.

[154] M. S. Johnstone and P. R. Wilson. The memory fragmentation problem: Solved? In *ACM SIGPLAN Notices*, volume 34, 26–36. ACM, 1998.

[155] A. Silberschatz, P. B. Galvin, and G. Gagne. *Operating System Concepts*, volume 4. Addison-Wesley Reading, 1998.

[156] R. R. Coifman and M. V. Wickerhauser. Entropy-based algorithms for best basis selection. *IEEE Transactions on Information Theory*, 38(2):713–718, 1992.

[157] P. Wright, M. C. Parker, and A. Lord. Simulation results of shannon entropy based flexgrid routing and spectrum assignment on a real network topology. In *IET Conference Proceedings*. The Institution of Engineering & Technology, 2013.

[158] D. Amar, E. L. Rouzic, N. Brochier, J. L. Auge, C. Lepers, and N. Perrot. Spectrum fragmentation issue in flexible optical networks: Analysis and good practices. *Photonic Network Communications*, 29(3):230–243, 2015.

[159] J. Kim, S. Yan, A. Fumagalli, E. Oki, and N. Yamanaka. An analytical model of spectrum fragmentation in a two-service elastic optical link. In *2015 IEEE Global Communications Conference (GLOBECOM)*, pages 1–6. IEEE, 2015.

[160] Y. Yu, J. Zhang, Y. Zhao, X. Cao, X. Lin, and W. Gu. The first single-link exact model for performance analysis of flexible grid wdm networks. In *National Fiber Optic Engineers Conference*, pages JW2A–68. Optical Society of America, 2013.

[161] B. C. Chatterjee, W. Fadini, and E. Oki. A spectrum allocation scheme based on first-last-exact fit policy for elastic optical networks. *Journal of Network and Computer Applications*, 68:164–172, 2016.

[162] W. Fadini, B. C. Chatterjee, and E. Oki. A subcarrier-slot partition scheme with first-last fit spectrum allocation for elastic optical networks. *Computer Networks*, 91:700–711, 2015.

[163] X. Chen, J. Li, P. Zhu, R. Tang, Z. Chen, and Y. He. Fragmentation-aware routing and spectrum allocation scheme based on distribution of traffic bandwidth in elastic optical networks. *Journal of Optical Communications and Networking*, 7(11):1064–1074, 2015.

[164] A. Pagès, J. Perelló, S. Spadaro, and J. Comellas. Optimal route, spectrum, and modulation level assignment in split-spectrum-enabled dynamic elastic optical networks. *Journal of Optical Communications and Networking*, 6(2):114–126, 2014.

[165] J. Comellas, X. Calzada, and G. Junyent. Efficient spectrum assignment in elastic optical networks. In *Transparent Optical Networks (ICTON), 2016 18th International Conference on*, 1–4. IEEE, 2016.

[166] Z. Zhu, X. Liu, Y. Wang, w. Lu, L. Gong, S. Yu, and N. Ansari. Impairment-and splitting-aware cloud-ready multicast provisioning in elastic optical networks. *IEEE/ACM Transactions on Networking*, 2016.

[167] Y. Qiu. An efficient spectrum assignment algorithm based on variable-grouping mechanism for flex-grid optical networks. *Optical Switching and Networking*, 24:39–46, 2017.

[168] S. Sugihara, Y. Hirota, S. Fujii, H. Tode, and T. Watanabe. Dynamic resource allocation for immediate and advance reservation in space-division-multiplexing-based elastic optical networks. *Journal of Optical Communications and Networking*, 9(3):183–197, 2017.

[169] S. K. Singh and A. Jukan. Efficient spectrum defragmentation with holding-time awareness in elastic optical networks. *Journal of Optical Communications and Networking*, 9(3):B78–B89, 2017.

[170] S. K. Singh, W. Bziuk, and A. Jukan. Analytical performance modeling of spectrum defragmentation in elastic optical link networks. *Optical Switching and Networking*, 24:25–38, 2017.

[171] M. Zhang, C. You, and Z. Zhu. On the parallelization of spectrum defragmentation reconfigurations in elastic optical networks. *IEEE/ACM Transactions on Networking*, 24(5):2819–2833, 2016.

[172] D. Medhi and K. Ramasamy. *Network routing: algorithms, protocols, and architectures*. Morgan Kaufmann, 2017.

[173] J. L. Gross and J. Yellen. *Graph Theory and its Applications*. CRC press, 2005.

[174] S. Ba, B. C. Chatterjee, S. Okamoto, N. Yamanaka, A. Fumagalli, and E. Oki. Route partitioning scheme for elastic optical networks with hitless defragmentation. *IEEE/OSA J. Opt. Commun. Netw.*, 8(6):356–370, 2016.

[175] P. M. Moura, N. L. D. Fonseca, and R. A. Scaraficci. Fragmentation aware routing and spectrum assignment algorithm. In *Communications (ICC), 2014 IEEE International Conference on*, 1137–1142. IEEE, 2014.

[176] P. M. Moura, R. A. Scaraficci, and N. L. D. Fonseca. Algorithm for energy efficient routing, modulation and spectrum assignment. In *Communications (ICC), 2015 IEEE International Conference on*, 5961–5966. IEEE, 2015.

[177] Z. Zhu, W. Lu, L. Zhang, and N. Ansari. Dynamic service provisioning in elastic optical networks with hybrid single-/multi-path routing. *Journal of Lightwave Technology*, 31(1):15–22, 2013.

[178] R. Zhu, Y. Zhao, H. Yang, X. Yu, J. Zhang, A. Yousefpour, N. Wang, and J. P. Jue. Dynamic time and spectrum fragmentation-aware service provisioning in elastic optical networks with multi-path routing. *Optical Fiber Technology*, 32:13–22, 2016.

[179] X. Lagrange and B. Jabbari. *Multiaccess, Mobility and Teletraffic for Wireless Communications*. Springer, 1999.

[180] A. Kadohata, A. Hirano, M. Fukutoku, T. Ohara, Y. Sone, and O. Ishida. Multi-layer greenfield re-grooming with wavelength defragmentation. *IEEE Communications Letters*, 16(4):530–532, 2012.

[181] T. Takagi, H. Hasegawa, K.I. Sato, Y. Sone, A. Hirano, and M. Jinno. Disruption minimized spectrum defragmentation in elastic optical path networks that adopt distance adaptive modulation. In *European Conference and Exposition on Optical Communications*, Mo–2. Optical Society of America, 2011.

[182] R. Proietti, C. Qin, B. Guan, Y. Yin, R. P. Scott, R. Yu, and S. J. B. Yoo. Rapid and complete hitless defragmentation method using a coherent RX LO with fast wavelength tracking in elastic optical networks. *Optics express*, 20(24):26958–26968, 2012.

[183] X. Wang, I. Kim, Q. Zhang, P. Palacharla, and M. Sekiya. A hitless defragmentation method for self-optimizing flexible grid optical networks. In *European Conference and Exhibition on Optical Communication*, P5–04. Optical Society of America, 2012.

[184] Y. Aoki, X. Wang, P. Palacharla, K. Sone, S. Oda, Takeshi Hoshida, Motoyoshi Sekiya, and Jens C Rasmussen. Dynamic and flexible photonic node architecture with shared universal transceivers supporting hitless defragmentation. In *European Conference and Exhibition on Optical Communication*, We–3. Optical Society of America, 2012.

[185] F. Cugini, M. Secondini, N. Sambo, G. Bottari, G. Bruno, P. Iovanna, and P. Castoldi. Push-pull technique for defragmentation in flexible optical networks. In *Optical Fiber Communication Conference*, JTh2A–40. Optical Society of America, 2012.

[186] M. Sekiya, Y. Aoki, and W. Xi. Photonic network defragmenttaion technology improving resource utilization during operation. *Fujitsu Sci. Tech. J*, 50(1):111, 2014.

[187] M. Zhang, Y. Yin, R. Proietti, Z. Zhu, and S. J. B. Yoo. Spectrum defragmentation algorithms for elastic optical networks using hitless spectrum retuning techniques. In *Optical Fiber Communication Conference*, OW3A–4. Optical Society of America, 2013.

[188] R. Wang and B. Mukherjee. Provisioning in elastic optical networks with non-disruptive defragmentation. *Journal of Lightwave Technology*, 31(15):2491–2500, 2013.

[189] W. Shi, Z. Zhu, M. Zhang, and N. Ansari. On the effect of bandwidth fragmentation on blocking probability in elastic optical networks. *IEEE Trans. on Commun.*, 61(7):2970–2978, 2013.

[190] F. Paolucci, A. Castro, F. Fresi, M. Imran, A. Giorgetti, B. B. Bhownik, G. Berrettini, G. Meloni, F. Cugini, L. Velasco, L. Poti and P. Castoldi. Active PCE demonstration performing elastic operations and hitless defragmentation in flexible grid optical networks. *Photonic Network Communications*, 29(1):57–66, 2015.

[191] F. Cugini, F. Paolucci, G. Berrettini, M. Secondini, F. Fresi, G. Meloni, N. Sambo, L. Potì, and P. Castoldi. Experimenting push-pull defragmentation in flexible optical networks with direct detection. In *Global Communications Conference (GLOBECOM), 2012 IEEE*, 2876–2881. IEEE, 2012.

[192] F. Cugini, F. Paolucci, G. Meloni, G. Berrettini, M. Secondini, F. Fresi, N. Sambo, L. Poti, and P. Castoldi. Push-pull defragmentation without traffic disruption in flexible grid optical networks. *Journal of Lightwave Technology*, 31(1):125–133, 2013.

[193] V. N. Rozental, S. M. Rossi, A. Chiuchiarelli, T. C. Lima, J. D. Reis, J. R. F. de Oliveira, and D. A. A. Mello. Gradual symbol rate switching for synchronous operation of flexible optical transceivers. *IEEE Photonics Technology Letters*, 28(4):505–508, 2016.

[194] S. Ba, B. C. Chatterjee, and E. Oki. Performance of route partitioning scheme for hitless defragmentation in elastic optical networks. *In proc. of IEEE ICNC*, 1–5. IEEE, 2017.

[195] B. Kozicki, H. Takara, Y. Tsukishima, T. Yoshimatsu, K. Yonenaga, and M. Jinno. Experimental demonstration of spectrum-sliced elastic optical path network (SLICE). *Opt. Express*, 18(21):22105–22118, Oct 2010.

[196] H. Takara, T. Goh, K. Shibahara, K. Yonenaga, S. Kawai, and M. Jinno. Experimental demonstration of 400 Gb/s multi-flow, multi-rate, multi-reach optical transmitter for efficient elastic spectral routing. In *European Conference and Exposition on Optical Communications*, Tu–5. Optical Society of America, 2011.

[197] L. Liu, R. Muñoz, R. Casellas, T. Tsuritani, R. Martínez, and I. Morita. OpenSlice: an OpenFlow-based control plane for spectrum sliced elastic optical path networks. *Optics Express*, 21(4):4194–4204, 2013.

[198] M. Channegowda, R. Nejabati, M. R. Fard, S. Peng, N. Amaya, G. Zervas, D. Simeonidou, R. Vilalta, R. Casellas, R. Martínez, R. Muoz, L. Liu, T. Tsuritani, I. Morita, A. Autenrieth, J. P. Elbers, P. Kostecki and P. Kaczmarek. Experimental demonstration of an OpenFlow based software-defined optical network employing packet, fixed and flexible DWDM grid technologies on an international multi-domain testbed. *Optics Express*, 21(5):5487–5498, 2013.

[199] B. Kozicki, H. Takara, Y. Tsukishima, T. Yoshimatsu, T. Kobayashi, K. Yonenaga, and M. Jinno. Optical path aggregation for 1-Tb/s transmission in spectrum-sliced elastic optical path network. *Photonics Technology Letters, IEEE*, 22(17):1315–1317, Sept 2010.

[200] R. Casellas, R. Muñoz, J. M. Fabrega, M. S. Moreolo, R. Martinez, L. Liu, T. Tsuritani, and I. Morita. Design and experimental validation of a GM-PLS/PCE control plane for elastic CO-OFDM optical networks. *Selected Areas in Communications, IEEE Journal on*, 31(1):49–61, 2013.

[201] F. Cugini, G. Meloni, F. Paolucci, N. Sambo, M. Secondini, L. Gerardi, L. Potì, and P. Castoldi. Demonstration of flexible optical network based on path computation element. *J. Lightwave Technol.*, 30(5):727–733, Mar 2012.

[202] J. Zhang, Y. Zhao, H. Yang, Y. Ji, H. Li, Y. Lin, G. Li, J. Han, Y. Lee, and T. Ma. First demonstration of enhanced software defined networking (eSDN) over elastic grid (eGrid) optical networks for data center service migration. In *National Fiber Optic Engineers Conference*, PDP5B–1. Optical Society of America, 2013.

[203] R. Proietti, R. Yu, K. Wen, Y. Yin, and S. J. B. Yoo. Quasi-hitless defragmentation technique in elastic optical networks by a coherent RX LO with fast TX wavelength tracking. In *Photonics in Switching Conference*, 1–3, 2012.

[204] S. Ma, C. Chen, S. Li, M. Zhang, S. Li, Y. Shao, Z. Zhu, L. Liu, and S. J. B. Yoo. Demonstration of online spectrum defragmentation enabled by OpenFlow in software-defined elastic optical networks. In *Optical Fiber Communication Conference*, W4A.2. Optical Society of America, 2014.

[205] R. Muñoz, R. Casellas, R. Martínez, L. Liu, T. Tsuritani, and I. Morita. Experimental evaluation of efficient routing and distributed spectrum

allocation algorithms for GMPLS elastic networks. *Optics Express*, 20(27):28532–28537, 2012.

[206] D. J. Geisler, R. Proietti, Y. Yin, R. P. Scott, X. Cai, N. Fontaine, L. Paraschis, O. Gerstel, and S. J. B. Yoo. The first testbed demonstration of a flexible bandwidth network with a real-time adaptive control plane. In *37th European Conference and Exposition on Optical Communications*, Th.13.K.2. Optical Society of America, 2011.

[207] S. Ma, C. Chen, S. Li, M. Zhang, S. Li, Y. Shao, Z. Zhu, L. Liu, and S. J. B. Yoo. Demonstration of online spectrum defragmentation enabled by OpenFlow in software-defined elastic optical networks. In *Optical Fiber Communication Conference*, W4A–2. Optical Society of America, 2014.

[208] P. Zhu, J. Li, D. Wu, Z. Wu, Y. Tian, Y. Chen, D. Ge, X. Chen, Z. Chen, and Y. He. Demonstration of elastic optical network node with defragmentation functionality and SDN control. In *Optical Fiber Communication Conference*, Th3I–3. Optical Society of America, 2016.

[209] F. Ji, X. Chen, W. Lu, J. J. P. C. Rodrigues, and Z. Zhu. Dynamic P-cycle protection in spectrum-sliced elastic optical networks. *Journal of Lightwave Technology*, 32(6):1190–1199, 2014.

[210] C. Wang, G. Shen, and S. K. Bose. Distance adaptive dynamic routing and spectrum allocation in elastic optical networks with shared backup path protection. *Journal of Lightwave Technology*, 33(14):2955–2964, 2015.

[211] G. Shen, H. Guo, and S. K. Bose. Survivable elastic optical networks: Survey and perspective. *Photonic Network Communications*, 31(1):71–87, 2016.

[212] S. Ba, B. C. Chatterjee, and E. Oki. Defragmentation scheme based on exchanging primary and backup paths in 1+1 path protected elastic optical networks. *IEEE/ACM Transactions on Networking*, 1–14, 2016, to appear.

[213] B. C. Chatterjee and E. Oki. Performance of hitless defragmentation scheme in quasi 1+ 1 path protected elastic optical networks. In *2018 20th International Conference on Transparent Optical Networks (ICTON)*, 1–4. IEEE, 2018.

[214] C. Wang, G. Shen, B. Chen, and L. Peng. Protection path-based hitless spectrum defragmentation in elastic optical networks: Shared backup path protection. In *Optical Fiber Communications Conference and Exhibition (OFC), 2015*, 1–3. IEEE, 2015.

[215] T. Sawa, F. He, T. Sato, B. C. Chatterjee, and E. Oki. Defragmentation using reroutable backup paths in toggled-based quasi 1+1 path protected

elastic optical networks. *IEICE Transactions on Communications*, E103-B(3), 2020.

[216] C. F. Hsu, H. C. Hu, H. F. Fu, J. J. Zheng, and S. X. Chen. Spectrum usage minimization for shared backup path protection in elastic optical networks. In *2019 International Conference on Computing, Networking and Communications (ICNC)*, 602–606. IEEE, 2019.

[217] D. S. Yadav, A. Chakraborty, and B. S. Manoj. A multi-backup path protection scheme for survivability in elastic optical networks. *Optical Fiber Technology*, 30:167–175, 2016.

[218] H. M. N. S. Oliveira and N. L. S. da Fonseca. Protection in elastic optical networks using failure-independent path protecting P-cycles. *Optical Switching and Networking*, 100535, 2019.

[219] D. R. Din. Survivable routing problem in EONs with fipp P-cycles protection. In *2017 IEEE International Symposium on Local and Metropolitan Area Networks (LANMAN)*, 1–2. IEEE, 2017.

[220] M. Klinkowski. An evolutionary algorithm approach for dedicated path protection problem in elastic optical networks. *Cybernetics and Systems*, 44(6-7):589–605, 2013.

[221] K. Walkowiak and M. Klinkowski. Shared backup path protection in elastic optical networks: Modeling and optimization. In *Design of Reliable Communication Networks (DRCN), 2013 9th International Conference on the*, 187–194. IEEE, 2013.

[222] Y. Wei, K. Xu, H. Zhao, and G. Shen. Applying p-cycle technique to elastic optical networks. In *Optical Network Design and Modeling, 2014 International Conference on*, 1–6. IEEE, 2014.

[223] H. M. N. S Oliveira and N. L. S. da Fonseca. Protection in elastic optical networks against up to two failures based FIPP P-cycle. In *Computer Networks and Distributed Systems (SBRC), 2014 Brazilian Symposium on*, 369–375. IEEE, 2014.

[224] S. Kosaka, H. Hasegawa, K. I. Sato, T. Tanaka, A. Hirano, and M. Jinno. Shared protected elastic optical path network design that applies iterative re-optimization based on resource utilization efficiency measures. In *European Conference and Exhibition on Optical Communication*, Tu–4. Optical Society of America, 2012.

[225] A. Tarhan and C. Cavdar. Shared path protection for distance adaptive elastic optical networks under dynamic traffic. In *2013 5th International Congress on Ultra Modern Telecommunications and Control Systems and Workshops (ICUMT)*, 62–67. IEEE, 2013.

[226] X. Shao, Y. K. Yeo, Z. Xu, X. Cheng, and L. Zhou. Shared-path protection in OFDM-based optical networks with elastic bandwidth allocation. In *Optical Fiber Communication Conference*, OTh4B–4, 2012.

[227] N. Kawase, R. Yamabayashi, M. Tomizawa, and Y. Uematsu. Route diversity with hitless path switching. *Electronics Letters*, 30(23):1962–1963, 1994.

[228] N. Kawase, Y. Yamabayashi, and Y. Uematsu. Hitless path switching apparatus and method, May 20 1997. US Patent 5,631,896.

[229] H. Ueda, T. Tsuboi, and H. Kasai. Hitless switching scheme for protected pon system. In *Global Telecommunications Conference, 2008. IEEE GLOBECOM 2008. IEEE*, 1–5. IEEE, 2008.

[230] T. Sawa, F. He, T. Sato, B. C. Chatterjee, and E. Oki. Defragmentation using reroutable backup paths in toggled 1+ 1 path protected elastic optical networks. In *2018 24th Asia-Pacific Conference on Communications (APCC)*, 422–427. IEEE, 2018.

[231] B. C. Chatterjee, T. Sato, and E. Oki. Recent research progress on spectrum management approaches in software-defined elastic optical networks. *Optical Switching and Networking*, 30:93–104, 2018.

[232] F. Bannour, S. Souihi, and A. Mellouk. Distributed SDN control: Survey, taxonomy and challenges. *IEEE Communications Surveys & Tutorials*, 20(1):333–354, 2018.

[233] B. C. Chatterjee, S. Ba, and E. Oki. Fragmentation problems and management approaches in elastic optical networks: a survey. *IEEE Communications Surveys & Tutorials*, 1–28, 2017.

[234] M. Channegowda, R. Nejabati, and D. Simeonidou. Software-defined optical networks technology and infrastructure: Enabling software-defined optical network operations. *Journal of Optical Communications and Networking*, 5(10):A274–A282, 2013.

[235] T. Szyrkowiec, A. Autenrieth, and W. Kellerer. Optical network models and their application to software-defined network management. *International Journal of Optics*, 2017.

[236] A. S. Thyagaturu, A. Mercian, M. P. McGarry, M. Reisslein, and W. Kellerer. Software defined optical networks (SDONs): A comprehensive survey. *IEEE Communications Surveys & Tutorials*, 18(4):2738–2786, 2016.

[237] D. B. Paredes, A. Beghelli, and A. Leiva. Network virtualization over elastic optical networks: A survey of allocation algorithms. In *Optical Fiber and Wireless Communications*. InTech, 2017.

[238] J. S. Dantas, D. Careglio, J. Perelló, R. M. Silveira, W. V. Ruggiero, and J. S. Pareta. Challenges and requirements of a control plane for elastic optical networks. *Computer Networks*, 72:156–171, 2014.

[239] I. Tomkos, S. Azodolmolky, J. S. Pareta, D. Careglio, and E. Palkopoulou. A tutorial on the flexible optical networking paradigm: State of the art, trends, and research challenges, 2014.

[240] F. S. Abkenar and A. G. Rahbar. Study and analysis of routing and spectrum allocation (RSA) and routing, modulation and spectrum allocation (RMSA) algorithms in elastic optical networks (EONs). *Optical Switching and Networking*, 23:5–39, 2017.

[241] M. Klinkowski, P. Lechowicz, and K. Walkowiak. Survey of resource allocation schemes and algorithms in spectrally-spatially flexible optical networking. *Optical Switching and Networking*, 2017.

[242] A. Alyatama, I. Alrashed, and A. Alhusaini. Adaptive routing and spectrum allocation in elastic optical networks. *Optical Switching and Networking*, 24:12–20, 2017.

[243] M. N. Dharmaweera, L. Yan, M. Karlsson, and E. Agrell. An impairment-aware resource allocation scheme for dynamic elastic optical networks. In *Optical Fiber Communication Conference*, pages Th2A–19. Optical Society of America, 2017.

[244] F. Köksal and C. Ersoy. Multicasting for all-optical multifiber networks. *Journal of Optical Networking*, 6(2):219–238, 2007.

[245] S. Sankaranarayanan and S. Subramaniam. Comprehensive performance modeling and analysis of multicasting in optical networks. *IEEE Journal on Selected Areas in Communications*, 21(9):1399–1413, 2003.

[246] A. K. Pradhan, B. C. Chatterjee, E. Oki, and T. De. Knapsack based multicast traffic grooming for optical networks. *Optical Switching and Networking*, 27:40–49, 2018.

[247] M. Moharrami, A. Fallahpour, H. Beyranvand, and J. A. Salehi. Resource allocation and multicast routing in elastic optical networks. *IEEE Transactions on Communications*, 65(5):2101–2113, 2017.

[248] J. R. de Almeida Amazonas, G. S. Boada, S. Ricciardi, and J. S. Pareta. Technical challenges and deployment perspectives of SDN based elastic

optical networks. In *Transparent Optical Networks (ICTON), 2016 18th International Conference on*, 1–5. IEEE, 2016.

[249] N. Yamanaka, K. Shiomoto, and E. Oki. *GMPLS technologies: Broadband backbone networks and systems*. CRC Press, 2005.

[250] L. Liu, D. Zhang, T. Tsuritani, R. Vilalta, R. Casellas, L. Hong, I. Morita, H. Guo, J. Wu, R. Martinez, and Ral Muoz. Field trial of an OpenFlow-based unified control plane for multilayer multigranularity optical switching networks. *Journal of Lightwave Technology*, 31(4):506–514, 2013.

[251] Z. Zhu, X. Chen, C. Chen, S. Ma, M. Zhang, L. Liu, and S. J. B. Yoo. OpenFlow-assisted online defragmentation in single-/multi-domain software-defined elastic optical networks [invited]. *Journal of Optical Communications and Networking*, 7(1):A7–A15, 2015.

[252] N. Cvijetic, A. Tanaka, P. N. Ji, K. Sethuraman, S. Murakami, and T. Wang. SDN and OpenFlow for dynamic flex-grid optical access and aggregation networks. *Journal of Lightwave Technology*, 32(4):864–870, 2014.

[253] M. Dallaglio, Q. P. Van, F. Boitier, C. Delezoide, D. Verchere, P. Layec, A. Dupas, N. Sambo, S. Bigo, and P. Castoldi. Demonstration of a SDN-based spectrum monitoring of elastic optical networks. In *Optical Fiber Communications Conference and Exhibition (OFC), 2017*, 1–2. IEEE, 2017.

[254] R. Casellas, R. Martínez, R. Muñoz, R. Vilalta, L. Liu, T. Tsuritani, and I. Morita. Control and management of flexi-grid optical networks with an integrated stateful path computation element and OpenFlow controller [invited]. *Journal of Optical Communications and Networking*, 5(10):A57–A65, 2013.

[255] L. Liu, Y. Yin, M. Xia, M. Shirazipour, Z. Zhu, R. Proietti, Q. Xu, S. Dahlfort, and S. J. B. Yoo. Software-defined fragmentation-aware elastic optical networks enabled by OpenFlow. In *IET Conference Proceedings*. The Institution of Engineering & Technology, 2013.

[256] H. C. Le, L. Liu, and S. J. B. Yoo. Distributed control plane with spectral fragmentation-aware RMSA and flexible reservation for dynamic multi-domain software-defined elastic optical networks. In *Optical Fiber Communication Conference*, Th2A–39. Optical Society of America, 2015.

[257] E. Oki. *Linear Programming and Algorithms for Communication Networks: A Practical Guide to Network Design, Control, and Management*. CRC Press, 2012.

[258] T. H. Cormen, C. E. Leiserson, R. L. Rivest, and C. Stein. *Introduction to Algorithms*. MIT Press, 2009.

[259] J. Kleinberg and E. Tardos. *Algorithm Design*. Pearson Education India, 2006.

[260] M. R. Garey and D. S. Johnson. *Computers and Intractability*, Volume 29. W H Freeman, New York, 2002.

[261] T. Baker, J. Gill, and R. Solovay. Relativizations of the P=?NP question. *SIAM Journal on Computing*, 4(4):431–442, 1975.

[262] R. M. Karp. Reducibility among combinatorial problems. In *Complexity of computer computations*, pages 85–103. Springer, 1972.

[263] B. Mukherjee. *Optical WDM networks*. Springer Science & Business Media, 2006.

[264] I. Chlamtac, A. Ganz, and G. Karmi. Lightpath communications: An approach to high bandwidth optical WAN's. *Communications, IEEE Transactions on*, 40(7):1171–1182, 1992.

[265] I. Tomkos, E. Palkopoulou, and M. Angelou. A survey of recent developments on flexible/elastic optical networking. In *Transparent Optical Networks (ICTON), 2012 14th International Conference on*, 1–6. IEEE, 2012.

[266] K. S. Abedin, T. F. Taunay, M. Fishteyn, D. J. DiGiovanni, V. R. Supradeepa, J. M. Fini, M. F. Yan, B. Zhu, E. M. Monberg, and F. V. Dimarcello. Cladding pumped erbium-doped multicore fiber amplifier. *Optics express*, 20(18):20191–20200, 2012.

[267] B. C. Chatterjee, Y. Jayabal, N. Yamanaka, S. Okamoto, A. Fumagalli, and E. Oki. Span power management scheme for rapid lightpath provisioning in multi-core fibre networks. *Electronics Letters*, 51(1):76–78, 2014.

[268] B. C. Chatterjee, F. He, E. Oki, A. Fumagalli, and N. Yamanaka. A span power management scheme for rapid lightpath provisioning and releasing in multi-core fiber networks. *IEEE/ACM Transactions on Networking*, 27(2):734–747, 2019.

[269] N. Sambo, F. Cugini, G. Bottari, P. Iovanna, and P. Castoldi. Distributed setup in optical networks with flexible grid. In *European Conference and Exposition on Optical Communications*, We–10. Optical Society of America, 2011.

[270] G. Shen and Q. Yang. From coarse grid to mini-grid to gridless: How much can gridless help contentionless? In *Optical Fiber Communication Conference*, OTuI3. Optical Society of America, 2011.

[271] R. Casellas, R. Muñoz, J. M. Fàbrega, M. Svaluto Moreolo, R. Martínez, L. Liu, T. Tsuritani, and I. Morita. GMPLS/PCE control of flexi-grid DWDM optical networks using CO-OFDM transmission [Invited]. *Optical Communications and Networking, IEEE/OSA Journal of*, 4(11):B1–B10, 2012.

[272] Open networking foundation (ONF) : Technical library. https://www.opennetworking.org/sdn-resources/technical-library. Last Accessed: 2015-01-30.

[273] A. Kretsis, P. Kokkinos, K. Christodoulopoulos, T. Varvarigou, and E. M. Varvarigos. Mantis: Cloud-based optical network planning and operation tool. *Computer Networks*, 77:153–168, 2015.

[274] H. A. Pereira, D. A. R. Chaves, C. J. A. Bastos-Filho, and J. F. Martins-Filho. OSNR model to consider physical layer impairments in transparent optical networks. *Photonic Network Communications*, 18(2):137–149, 2009.

[275] B. C. Chatterjee, N. Sarma, and P. P. Sahu. A QoS-aware wavelength assignment scheme for optical networks. *Optik-International Journal for Light and Electron Optics*, 124(20):4498–4501, 2013.

[276] C. Wang, G. Shen, and L. Peng. Protection path-based hitless spectrum defragmentation for elastic optical networks: 1+ 1 path protection. In *Asia Communications and Photonics Conference*, AF3E–3. Optical Society of America, 2014.

[277] Y. Kishi, N. Kitsuwan, H. Ito, B. C. Chatterjee, and E. Oki. Modulation-adaptive link-disjoint path selection model for 1+1 protected elastic optical networks. *IEEE Access*, 7(1):25422–25437, 2019.

Index

Milton Keynes UK
Ingram Content Group UK Ltd.
UKHW051015071024
449327UK00012B/256

9 780367 510213